Climate Under Cover

Climate Under Cover

by

Tadashi Takakura

Nagasaki University,
College of Environmental Studies, Japan

and

Wei Fang

National Taiwan University,
Department of Bio-Industrial Mechatronics Engineering, Taiwan

KLUWER ACADEMIC PUBLISHERS
DORDRECHT / BOSTON / LONDON

A C.I.P. Catalogue record for this book is available from the Library of Congress.

ISBN 1-4020-0845-7 (HB)
ISBN 1-4020-0846-5 (PB)

Published by Kluwer Academic Publishers,
P.O. Box 17, 3300 AA Dordrecht, The Netherlands.

Sold and distributed in North, Central and South America
by Kluwer Academic Publishers,
101 Philip Drive, Norwell, MA 02061, U.S.A.

In all other countries, sold and distributed
by Kluwer Academic Publishers,
P.O. Box 322, 3300 AH Dordrecht, The Netherlands.

Printed on acid-free paper

Printed in the Netherlands.

CONTENTS

vi

vii

PREFACE OF 1st EDITION

After plastic films were invented, covering technique became popular for crop protection. Mulching, row covers, floating mulch and greenhouses are classified differently, but in principle they are more or less the same technique. As far as the physical environment for plant growth is concerned, the main sub-systems are the soil layer and the boundary air layer above it. Covering the soil surface with film changes these environments drastically. Air space between the cover and the soil surface is very limited in mulching. If this space is enlarged and crops are grown under the film, the floating mulch, row cover and plastic house systems can be utilized.

Therefore, insofar as the physical environment itself is concerned, air space between the cover and the soil surface is the key to the classification. The same mechanism governs environments produced by the various covers. Therefore, in this book we will analyze all of the different environments from mulching to greenhouses. In order to analyze the physical environment, the relationship between plants and environment is of course another important topic.

Stress is placed on the link between quantitative phenomena and qualitative analyses, although not all simulation results are verified experimentally. Selection of adequate parameter values is for verification of the simulation. Most phenomena involved are non-linear and non-steady state. For this reason, the approach which is called System Dynamics is used and simulation models developed in a simulation language CSMP (Continuous System Modeling Program) are fully used. Simulation languages for continuous systems work on the same principles as analog computers and are problem-oriented. It is very easy to understand. There are several languages, such as micro-CSMP, ACSL, SYSL, and PCSMP which are run on computers ranging from mainframes to IBM PCs, and each is slightly different. The first three are available commercially. The last, PCSMP (CSMP for IBM PCs) has been developed by the Department of Theoretical Production Ecology, Agricultural University, Wageningen, the Netherlands. These simulation languages are much more powerful than BASIC or FORTRAN; they can be run on IBM PCs, and they satisfy engineers' nostalgia for analog computers.

This book was written for a computer simulation class at the graduate level at the University of Tokyo. The author assumes that readers have some basic background in differential equations, numerical analysis, and FORTRAN programming. He will be more than satisfied if the digital simulation technique using CSMP described in this book becomes widely used as a research tool as well as an educational tool in student laboratories. All models in this book are written in micro-CSMP and are available on request (a request card is attached to this book).

There are already some good English books for students majoring in agricultural structures and environments. However, they are mostly for animal environments and as far as the author knows simulation technique has not yet been well documented in the area of plant growth, although it is a very powerful methodology. The author started greenhouse simulation as a graduate student, using FORTRAN on mainframes. He became acquainted with digital simulation techniques during his post-doctoral study (1968-1969) at the University of Minnesota after having some difficulty in simulating his models using both hybrid and analog computers. Dr. L. L. Boyd, head of the department at that time, offered the author a chance to study there, and Dr. K. A. Jordan taught him how to use hybrid and analog computes. The first model developed at that time was made using FORTRAN (Takakura et al., 1971), and it has been referred to by many researchers on greenhouse environments since then. As is indicated in Takakura and Jordan (1970), the second model was in MIMIC (a similar simulation language as CSMP), which was only available at Minnesota. After the author started simulation work using CSMP in Japan, he had a chance to work in the Department of Theoretical Production Ecology (which was headed by Prof. Dr. C. T. de Wit at that time) at Agricultural University, Wageningen in the Netherlands in 1973, and also had an excellent opportunity to discuss digital simulation in CSMP with other members of his department. They cooperated with the staff of the computer science department at the same university to develop PCSMP later.

The writing of this book began when the author visited the Department of Biological and Agricultural Engineering, Rutgers University in 1987. A simulation model was developed for floor heating greenhouses with Prof. W.J. Roberts, the chairman of the department, and his young staff members. In the summer 1991, the author offered a one-week short course at Rutgers using the first draft of this book. All 15 attendants were eagerly discussed the contents of this book and gave many suggestions for its completion; especially helpful was Dr. Wei Fang, National Taiwan University. Dr. Ken Jordan, now in the University of Arizona ran the most of the models in ACSL and gave many suggestions. Nancy K. Okamura, research fellow in our department and Susan Schmidt, University of Tokyo Press read the manuscript carefully and made corrections. There are many others who contributed by giving suggestions and preparing the manuscript. This book would not have been possible without their help. The author would like to thank all who have contributed in many ways. Last but not the least, he would like to express his appreciation to Dr. Luweis, Editor in Chief, Kluwer Academic Publishers, who agreed to publish this book.

In Tokyo
October 1992

Tadashi Takakura

PREFACE OF 2nd EDITION

Since the first edition was published almost ten years ago and computer software has changed drastically. As far as the authors know, any of CSMP group is not supported for recent WINDOWS system. On the other hand, general mathematical software such as MATLAB has been taught in many engineering courses. Since we left Rutgers, we have been working at different institutions in different countries but in the same research area. Recent computer software development has forced us to learn mathematical software and both of us selected MATLAB. We have met at several occasions such as international symposium and one time we discussed to renew the first edition by using some new language which is more powerful and suitable in the present WINDOWS environment. This is the main reason we selected MATLAB. It has also very powerful toolboxes such as SIMULINK, which is very suitable for dynamic simulation. First we planned to involve all models in MATLAB and SIMULINK, but because of the limit of the volume, we have decided to have MATLAB models with one SIMULINK model as an example. It is not difficult to develop SIMULINK models if the readers understand MATLAB models. All models in the first edition have been converted in MATLAB with its explanation by Dr. Wei Fang. All models listed in this book can be downloaded from Dr. Wei Fang's home page. The URLs of the web sites are as follow:

> http://ecaaser3.ecaa.ntu.edu.tw/weifang/cuc/index.htm
> http://ecaaser5.ecaa.ntu.edu.tw/weifang/cuc/index.htm

Both of us hope all readers can run all models in the new computer environment.

In Nagasaki and Taipei
January 2002

Tadashi Takakura and Wei Fang

CHAPTER 1

INTRODUCTION

1.1. INTRODUCTION

Plastic covering, either framed or floating, is now used worldwide to protect crops from unfavorable growing conditions, such as severe weather and insects and birds.

Protected cultivation in the broad sense, including mulching, has been widely spread by the innovation of plastic films. Paper, straw, and glass were the main materials used before the era of plastics. Utilization of plastics in agriculture started in the developed countries and is now spreading to the developing countries.

Early utilization of plastic was in cold regions, and plastic was mainly used for protection from the cold. Now plastic is used also for protection from wind, insects and diseases. The use of covering techniques started with a simple system such as mulching, then row covers and small tunnels were developed, and finally plastic houses. Floating mulch was an exception to this sequence: it was introduced rather recently, although it is a simple structure. New development of functional and inexpensive films triggered widespread use of floating mulch.

Table 1.1. The use of plastic mulch in the world (after Jouët, 2001).

	1991	1999
World		12,130,000 (ha)
Africa and Middle East		80,000
Egypt	7,000	30,000
Israel	4,000	26,000
America		200,000
U.S.A.	20,000	75,000
Asia		9,760,000
China	1,400,000	9,600,000
Japan	150,000	160,000
Europe		450,000
Spain	100,000	150,000
France	100,000	100,000
Italy	50,000	75,000

Table 1.1 shows the use of plastic mulch in the world. Wide use in Asia especially drastic increase in China in the last ten years is apparent. China had the largest area of plastic mulch in the world followed by Japan, Spain, and France in 1999. The European region was second next to Asia, and Spain, France and Italy were the top three countries in Europe.

1

Table 1.2. The area of low tunnels and floating mulch (after Jouët, 2001; original data was modified by CIPA data, 2000).

Low tunnels	1991	1999
World		882,000(ha)
Africa		80,000
Egypt	15,000	50,000
America		30,000
U.S.A.	5,000	15,000
Asia		170,000
China	20,000	85,000
Japan	53,624	56,500
Europe		90,000
Italy	100,000	24,000
Spain	100,000	17,500
France	50,000	16,000
Floating mulch	**1991**	**1999**
World		86,000(ha)
Africa and Middle East		6,000
Egypt		5,000
America		6,000
U.S.A.	1,500	5,000
Asia		14,000
China	1,000	8,000
Japan	4,000	6,000
Europe		60,000
United Kingdom	9,000	12,000
France	8,000	10,000
Italy	3,000	10,000

Table 1.2 lists countries where the area of low tunnels and floating mulch is more than 15,000 ha. Since construction and dismantling of these facilities occur frequently statistical data are less reliable than those for more stable structures such as glasshouses. Again, China had the largest area of low tunnels in the world and Japan was the next. Floating mulch was most widespread in Europe, followed by Asia, America, and Africa and the Middle East.

These data are all from the Comité International des Plastiques en Agriculture (CIPA) in France, which is one of the organizations that publish this kind of worldwide statistical data, but it is not a governmental body and has some difficulty in collecting and updating these data.

The area of plastic greenhouses in the main horticultural countries is shown in Table 1.3 from largest to smallest for areas over 2,000 ha in each region. More than half of the world area of plastic greenhouses is in Asia and again China has the largest area. In most countries, greenhouses are made of plastic and glass; the majority are plastic, although, as is well known, almost all greenhouses are glass in the Netherlands where the greenhouse area is over 8,000 ha. Glasshouses and rigid-plastic houses are longer-life structures, and therefore are mostly located in cold regions where these structures can be used throughout the year. In Japan, year-round use of greenhouses is becoming predominant, but in moderate and warm climate regions they are still provisional and are only used in winter. Double-cropping systems consisting of rice cultivation in open fields and strawberry cultivation in plastic houses on same fields, are common in some regions in Japan.

1.2. EUROPE AND AMERICA

There is no such steep expansion of protected cultivation area in Europe and America as in Asia, but the total area is not small, and a stable situation exists. Protected cultivation into Europe is classified in three types: Fully automated greenhouses, simple plastic houses, and large-scale greenhouses associated with energy plants. A close look at protected cultivation in northwestern Europe shows that new systems have gradually been installed. Climatic control by digital computers, hydroponic systems, and inter- transportation systems (which can be called agri-robot systems) are spreading in northwestern European countries. In the Mediterranean countries, simple plastic houses are predominant. In the eastern European countries, large-scale glasshouses with energy supplied from nearby power plants are very common; they are normally operated cooperatively, although levels of mechanization are not high.

In the United States, greenhouses are used for farming in long-term structures on the one hand, while row covers and floating mulches are becoming popular on small farms (Schales, 1987; Wells, 1988). Some countries in Central and South America are becoming exporting countries of flowers produced in protected cultivation.

1.3. ASIA

Plastic films are widely used in Korea and China. In Korea, plastic houses are also used for temporary protected fruit culture 1) to control low temperature; 2) to avoid diseases and damage by insects and birds; 3) to shut out strong rain and wind; and 4) to improve the fruit quality (Park, 1988). A similar approach is becoming used in more tropical Asian countries such as Thailand and the Philippines (Bualek, 1988; Bautista, 1988).

Table 1.3.   The area of plastic greenhouses in the main horticultural countries   (after
Jouët, 2001; the original data was modified by CIPA data, 2000).

	1991	1999
World		682,050 (ha)
Africa and Middle East		55,000
Turkey		14,000
Morocco		10,000
Israel	1,500	5,200
Algeria	4,802	5,005
Rep. of South Africa		2,500
Syria		2,000
America		22,350
U.S.A.	2,850	9,250
Colombia		4,500
Ecuador		2,700
Asia		450,000
China	200,000	380,000
Japan	45,033	51,042
Korea		2,200
Europe		180,000
Italy	65,000	61,900
Spain	35,000	51,000
France	9,000	9,200
Hungary	4,000	6,500
Serbia		5,040
Czech and Slovak Rep.		4,300
Russia	4,850	3,250
Greece		3,000
Portugal		2,700
United Kingdom	1,000	2,500
Poland		2,000

1.4. AFRICA AND THE MIDDLE EAST

The weather in the Mediterranean region is relatively mild.   Turkey, Morocco, Israel, and Algeria are all in this region and rather simple structures are predominant in these countries.

1.5. JAPAN

In Japan, primitive methods using oil paper and straw mats to protect crops from the severe natural environment were used as long ago as the early 1600s. Widespread implementation of protected cultivation in Japan began in 1951 after the introduction of Polyvinyl Chloride (PVC) film, and the benefits of its application to agriculture were quickly appreciated. Paper covered tunnels were rapidly replaced by PVC-film covered ones. Traditional wooden or bamboo frames were replaced with steel first and then with aluminum and plastics in some cases. Mulching with plastics began with PVC film, but later shifted to Polyethylene (PE) after its introduction (Nishi, 1986).

Protected cultivation will be one of the promising ways of supplying food under unfavorable environmental conditions in Japan if the energy problems can be solved.

The past and present situation in Japanese protected cultivation is summarized in Table 1.4 and Fig. 1.1. (The data in the figure are only for greenhouses.) This figure shows rather steady expansion of protected cultivation area has continued for years along with the development of new covering materials, but it is clear that the expansion rate of plastic houses flattened during the steep oil price increases in the 1970's. One of the particular trends in Japan was that the area for pomiculture expanded along with that for floriculture, and in the 1980's the pomiculture area was at one time larger than the area for floriculture. The rapid growth of rain shelter area in the 1980's and 1990's is apparent after rain shelters were introduced in 1983 (see Fig. 1.1).

Statistics on the use of agricultural plastics in recent years are summarized in Table 1.4. The area of mulching is predominant. Tunnels follow, with more than half using PVC. The most dominant use of PVC is, however, in greenhouses. Direct film covering of crops without the use of frames --floating mulch -- is increasing dramatically because of its higher yields and lower costs. Direct net covering is also consistently being used to prevent crops from damage caused by insects, birds and severe winds. In tropical regions, film covering is used not only for rain shelters but also for weed and soil moisture control. The use of plastic covering in agriculture is based on the light transmission and thermal properties of plastic films, and their effects in the crop environment.

Table 1.4. Estimation of the world consumption of plastic used in agricultural production (after Jouët, 2001).

	1985	1991	1999
Tunnels	88,000	122,000	168,000 (t)
Mulch	270,000	370,000	650,000
Floating mulch	22,500	27,000	40,000
Greenhouses	180,000	350,000	450,000
Silages	140,000	265,000	540,000
PP twine for hay and straw	100,000	140,000	204,000
Hydroponic systems	5,000	10,000	20,000
Micro-irrigation	260,000	325,000	625,000
Others (nets, plastic bags, except fertilizer bags)	80,000	130,000	150,000
Total	1,145,500	1,759,000	2,847,000

1.5.1. COVERING MATERIALS: PE OR PVC?

Covering materials are more or less common to all systems, from mulching to greenhouses. Glass has been the traditional and ideal covering material for many years, and it still is, for facilities designed for year-round use. When farmers insist on investment in a shorter period or use a facility for only a certain part of the year, soft films are used. The three covering films most often used are PVC, PE and Ethylene Vinyl Acetate (EVA). PVC is most opaque to long wave radiation (second only to glass); next most opaque is EVA, and then PE at the same thickness.

PE is most transparent to long wave radiation; often the inside air temperature of a PE house can be lower than the outside due to long wave radiation loss on calm, clear winter nights. The few degrees difference in indoor temperature between a PVC and a PE house is critical if the house is not heated and the temperature is around freezing. Now that IR-resistant PE and EVA have been developed, the difference can be minimized. The long wave transmissivity characteristics of IR-resistant PE are not reported, but the order of thermal superiority would be PVC first, then IR-resistant PE and EVA at almost the same level (Grafiadellis, 1985). The situation is not so critical in heated greenhouses.

In the 1980s, a report indicated that evaporation of Dibutyl Phthalate, which is used as a plasticizer, from PVC films and pipes harms plants inside the house. In Japan, this substance was replaced almost 35 years ago with Di-Ocpyl Phthalate (Di-Ethyl Hexyl Phthalate), a harmless chemical. Deterioration of films is also slower today because of the longer life of plasticizers in films. At the International Seminar on Agricultural Plastics in Korea in 1988, a scientist reported on discharge of chloride gas from PVC film greenhouses. The gas can be generated if the film is burned, but not if the film is at a temperature of less than around 180°C. Not only these chemical properties of the films but also their physical properties entirely depend upon how they are made in each country. In Japan, Japanese Industrial Standards (JIS), similar to DIN in Germany, are used to judge the products. In North America each manufacturer reportedly sets its own standards (Blom and Ingratta, 1985). It is rather difficult to discuss the issue with such different products being produced under the same name.

Rigid polyester rigid film, which has much greater durability as well as transmissivity than PE and PVC, is becoming popular in Japan, although the price is rather high. Its lifetime is said to be 5 to 7 years.

Environmental conditions change the situation, of course. If condensation occurs, the inside surface of PE film becomes more opaque to long wave radiation.

Aging and weathering effects on films are also important factors to consider in selecting films, but this topic is beyond the scope of this book.

1.5.2. GREENHOUSE PRODUCTION IN JAPAN

In Japan, although the area of greenhouses is less than 1% of the total arable land, Table 1.5 shows the extent to which the country depends upon production in greenhouses for its main vegetables. Significant proportions of vegetables such as tomatoes, cucumbers, green peppers and strawberries are produced under protected cultivation.

In recent years, rain shelter greenhouses have been increasingly used in fruit and leaf vegetable production. Because of this recent increase, the area of rain shelter greenhouses is normally separated statistically from the other greenhouse area. However in 1989, the rain shelter area was approximately 10,000 ha, a 10% increase from the preceding year (see Fig. 1.1). Tomatoes and spinach accounted for more than 85% of the rain shelter greenhouse area. The normal open-field growing season is March or April through October or November, when outside temperature conditions are favorable for open field cultivation. With the use of greenhouses, the production area can then be from 200 to 300 m higher in altitude than traditional areas. In summer, day temperature is slightly higher inside, with no significant difference in night temperature. Fruit injury and disease infection due to rain have been drastically reduced: for example, tomatoes are substantially protected from bacterial canker and spinach from downy mildew.

1.5.3. FLOATING MULCH IN JAPAN

Floating mulch was first introduced in Okinawa (southernmost island in Japan) in the 1950s, in order to protect crops against severe weather conditions in summer. It has now become a common method in Okinawa, and has been adopted in other areas of Japan.

Floating mulch has various advantages: 1) reduction in damage caused by typhoons, 2) protection from severe swings in weather, and 3) reduction in damage by birds and insects. Materials for films are the non-woven fabric of Polyethylene Vinyl Alcohol (EVOH), PE, polyester and Polypropylene (PP), and materials for cheesecloth and net are polyester and PVA. The physical properties of these materials are listed in Table 1.6.

Table 1.5. The ratio of vegetable production in greenhouses to total production from 1975 to 1998 in Japan (after JGHA, 2001).

Crop	1975	1979	1983	1990	1998
Eggplant	32	35	37	38	34 (%)
Tomatoes	34	40	54	68	71
Cucumbers	49	53	57	62	65
Pumpkins*	31	37	37	36	39
Green peppers	60	66	65	68	64
Strawberries	81	85	89	95	99
Watermelons*	63	73	77	83	94
Lettuce*	28	30	32	32	35

* Tunnel cultivation is included.

Table 1.6. Main physical properties of materials available in Japan for floating mulches (after Takakura, 1988).

Materials	Light trans-missivity(%)	Emissivity (%)*	Weight (g/m^2)	Price $(\$/m^2)$
Non-woven				
EVOH	95	81	40	0.6 - 0.8
PE	90	13	35	0.4 - 0.5
Polyester	70 - 90	53	15 - 30	0.2 - 0.4
PP	80 - 90	16	20 - 25	0.2 - 0.3
Cheesecloth				
Polyester	70 - 90	53	30 - 50	0.8 - 1.0

* Emissivity of crude materials

1.5.4. PLASTIC WASTES

Waste treatment of used plastics has been one of the biggest problems since the amount of plastic consumption has increased drastically in recent years. In 1985, the total amount of waste exceeded 165,892 tons: 68.7 % was from vegetable cultivation and 12.0 % from field crop cultivation; 55.1 % was PVC and 38.2 % was PE.

Since 1970, plastic wastes have been required to be treated under industrial waste regulations; all users, mostly growers, must be responsible for treating the wastes they generate without producing any air and water pollution. It is illegal to throw away plastic wastes untreated because they form obstacles in rivers and other public places.

The methods of treating plastic wastes in Japan are 1) recycling, 2) burial, and 3) incineration (Table 1.7).

Recycling

Five types of recycling are used. a) Pellets and fluffs are generated from waste plastics. Collected waste plastics, mostly PVC, are first graded, and foreign materials are removed. After a rough crush, they are washed with water, and fine crush and drying make them final products. The recovery ratio from used materials is approximately 50% for PVC. The products are half-materials for plastic tiles, mats, sandals and fillers. b) Collected waste plastics, either PVC or PE, are crushed and then melted without washing. Plastic are extruded to make the final plastic products. c) Collected waste PE is crushed and mixed with sawdust or rice hulls to make solid fuels. Their calorific values vary from 5,630 to 10,050 kcal/kg, which is equivalent to that of coals and cokes. d) Waste PVC and PE can be treated to make hydrophobic materials for under-draining, (*i.e.*, plastic trays and pipes). e) Oil or gas can be regenerated by pyrolysis of waste PE.

Table 1. 7. Treatment of waste plastics in Japan (after JGHA, 2001).

Year	1983	1987	1991	1995	1999
Recycling	31,058	33,134	42,565	43,238	49,812
Burial	26,832	35,828	41,958	43,238	49,812
Incineration	64,620	75,801	76,090	72,448	31,423
Other*	35,390	29,946	23,303	22,147	35,319
Total (t)	157,900	174,709	183,916	190,515	178,887

* includes collection by companies.

Burial

Waste plastics that are not suitable for recycling must be buried according to the law of Japan, which also regulate the place and method of burial and pre-treatment.

Incineration

Incineration is also regulated by law. It is not recommended, but incineration of up to 100 kg/day at an authorized facility is allowed. More detailed and thorough approaches will be needed both technically and administratively.

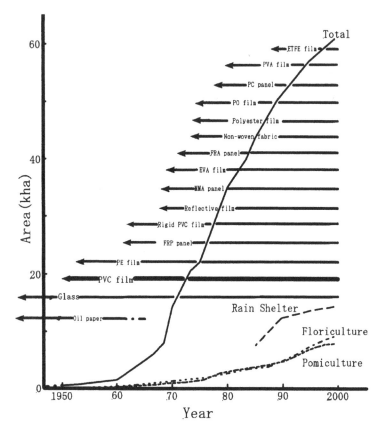

Figure 1.1. Trends in greenhouse development and covering materials in use in Japan (after JGHA, 2001).

PROBLEMS

1. List the concrete advantages of the usage of plastic houses over open-field cultivation.

2. Why is air temperature in PE houses in general lower on cold nights than that in unheated PVC houses?

3. List the concrete advantages of floating mulches.

4. Describe briefly the recycling methods for waste plastics.

CHAPTER 2

DEFINITION OF COVERING AND PROPERTIES OF COVERING MATERIALS

2.1. INTRODUCTION

There are many kinds of covering materials and techniques available and used in practical agriculture. It is rather difficult to distinguish one technique from another in some cases because the principle involved is the same. The following terms are used for these systems: mulchings, row covers (tunnels), floating mulches (floating row cover), rain shelters, and unheated greenhouses.

There is a definite difference between mulching and the other techniques. In mulching, plants grow over the cover; in the others, they grow under. Some techniques are combined in use. Figure 2.1 shows pictorial definitions of these techniques.

2.2. MULCHING

Several kinds of materials -- not only plastics but also organic materials -- are used for mulching. In mulching, the soil surface where crops are growing is covered directly by these materials, with a very thin air space between the cover and the soil surface. In order to improve the thermal environment of the soil, film mulching is widely used both in open fields and in greenhouses.

The soil surface of each row is covered by a sheet of film with many holes. Plants are sown or transplanted into each hole, and grow over the cover. Since most of the soil surface is covered by film, evaporation from the surface is prevented. It is clear that the soil is kept warmer than it would be without mulching.

Table 2. 1.   *Effect of plastic mulch on marketable yields of early cucumbers and muskmelons in summer of 1965 (after Jensen, 1988).*

Treatment	Cucumber (kg/ha)	Muskmelon (kg/ha)
Clear plastic	37,569	13,362
Black plastic	33,336	9,786
No plastic	26,043	911

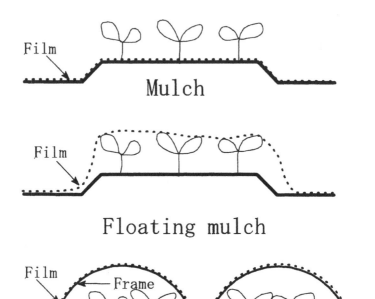

Mulch

Floating mulch

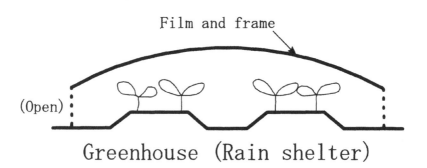

Small tunnel or Row cover

Greenhouse (Rain shelter)

Figure 2.1. Covering techniques.

The color of film used for mulching varies, from black to white to clear.   Each has its own specific characteristics, (details are discussed in Chapter 6).

Transparent film creates the highest soil temperatures because solar radiation penetrates the soil directly. With black film, the temperature in the soil is slightly lower than that in transparent-film covered soil; but weeds are killed in the absence of light, and the soil environment and crop yields are improved.   Some examples of crop yields are shown in Table 2.1. Covering with reflective film reduces the temperature of the soil during the hot summer months.   Water conditions and virus infections in soils are also largely affected by coverings.   Greenhouse floor soils are often covered with white-colored film to protect rockwool systems from soil diseases.   The use of mulching not only improves the temperature environment but also achieves various other goals, such as moisture control in soil, sterilization for weed control and soilborne diseases, and enhanced light reflection for early production.

2.3. ROW COVERS (TUNNELS)

Row covers or small tunnels are film-covered shelters with small frames; normally, farmers cannot work inside them. Size is the main difference between row covers and unheated plastic greenhouses.   For row covers, a sheet of film is placed over the plants in a row with an arch-shaped frame for support.   The shape of the frame can vary, but one unit of frame fits only one row, as is shown in Fig. 2.1, and the sheet of film covers more than one row.   This can be thought of as a kind of floating row cover or floating cover, which usually does not use any frame.

The thermal properties of films are very important in row covers because the rows are not heated.   PVC film has a very high emissivity for long-wave radiation (similar to glass), which creates slightly higher air temperatures in the sheltered space.   This improvement in thermal environment completely outweighs the price advantage of the less expensive PE film.   The big advantage of row covers is that the films can be removed during the summer season.   On the other hand, thermal insulation may also be improved by multi-layer coverings.   Inside row covers are used especially for growing seedlings and can be removed when the crops grow to a certain height.   Opaque sheets can also be applied at night.   Floating mulches have been used recently as an alternative to inside row covers.   More than 90% of heated greenhouses in Japan have at least one layer of thermal screen, which is retractable.

2.4. FLOATING MULCHES (FLOATING ROW COVER)

Floating mulch is the technique of covering crops with film without using any frame, although tape or wire over the film is sometimes used to fasten and stabilize the film against wind.

In most cases, the plants are covered by sheets of film or net without frames.   If frames are used it is rather difficult to separate this method from row covering, but floating methods have been developed recently for wider purposes using various coverings such as nets for wind protection.   If introduction of frames seems likely to improve the situation, frames may be added.

Floating mulch has various advantages: 1) reduction in damage by typhoons; 2) protection from severe swings in weather; and 3) reduction in damage by birds and insects.

2.5. RAIN SHELTERS

It can be said that rain shelters are unheated greenhouses with wider side openings. They are used year-round but particularly in the warmer season for protection from rain.

2.6. UNHEATED GREENHOUSES

Greenhouses without any heating or cooling facility are basic structures for environmental control and can achieve a better temperature environment in the passive sense than rain shelters.   The basic behaviour of greenhouses cannot be changed by the addition of sophisticated environmental control facilities.   In cool regions, greenhouses and row covers are often used in combination.   Solar energy is stored during the daytime; therefore, the inside temperature is normally higher than outside during the nighttimes, which contributes to plant growth.

2.7. COVERING MATERIALS

Various kinds of coverings are used for each method because each method has its own specific characteristics (see Fig. 2.2).   In the descriptions that follow, the differences between production methods and detailed characteristic structures, such as low and high density and linear low density, are not mentioned, because durability and economic aspects are beyond the scope of this book.

There are mainly two ways to classify films: One is based on their components and production method, and the other is based on their physical properties such as reflectivity and permeability.   In classification of the components, the use of abbreviations is popular. PVC (Polyvinyl chloride) and PE (Polyethylene) films are two predominant films used for almost all methods.   Other films include EVA (Ethylene vinyl acetate), EVOH (Polyethylene vinyl alcohol), PP (Polypropylene), PETP (Polyethylene telephtalate), FRP (Fiber reinforced plastic), FRA (Fiber reinforced acrylic), PMMA (Polymethyl methacrylate) and PC (Polycarbonate).

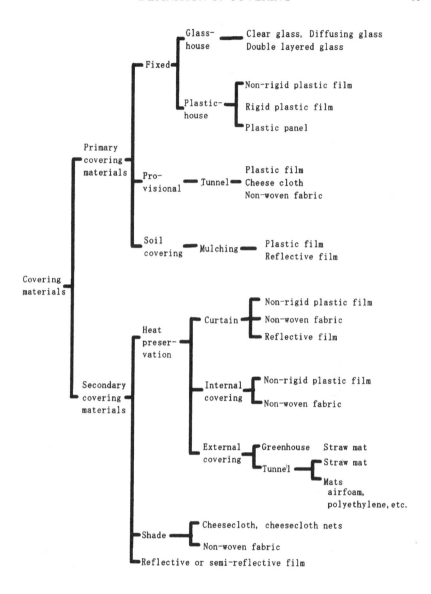

Figure 2.2. Classification of covering materials (JGHA, 1986).

Films for agriculture are mostly made from PVC, EVA, EVOH and PP. Rigid films (which are not necessarily thicker but are harder in nature) are made from PVC and PETP. The difference between soft and hard films that are both made from PVC is the plasticizer content. Films composed of less than 15% plasticizer are soft in general. Rigid panels of thickness 0.45 to 1.7 mm are made from FRP, PRA, PMMA and PC. Not only flat panels but also corrugated and double sheet panels are available.

Reflective films, non-woven materials and nets are some other types of covering materials. Anti-droplet, anti-dust and water vapor permeability are their most important properties. Infra-red transparency is again an important draw back resulting in the poor ability of these materials to maintain higher temperatures inside. Some improvement has been noted for PE, which is originally mostly transparent to infra-red radiation.

The general characteristics of main covering materials are summarized in Table 2.2. The workability of films is in general good. Stickiness, an undesirable property as it is considered to accelerate dust accumulation, is high for PVC. Adhesion, a desirable property for most uses, is also the highest for PVC. Transparency is more or less the same for all materials, but is high in PVC and glass. Heat resistance is high in PVC, FRP and glass. Weather tolerance is rather poor in PE and EVA.

Cost is high for rigid covering materials such as FRP and glass; PVC is the most expensive among the films.

Table 2. 2.   Characteristics of covering materials.

		PVC	PE	EVA	FRP	Glass
	Transparency	◎	○	○	△	◎
	Strength	○	△	△	◎	○
Physical	Thermal resistance	○	X	△	○	◎
Properties	Dust prevention	X	○	○	○	○
	Anti-droplets	○	X	△	△	○
	Weather tolerance	○	△	△	○	◎
Workability		○	○	○	X	X
Cost		△	◎	○	X	X

◎ excellent, ○ good, △ fair, and X poor.

Other important characteristics are those related to thermal radiation characteristics such as transmissivity, absorptivity and reflectivity for long-wave radiation. The characteristics are listed in Table 2.3. They are monochromatic and change according to radiation wavelength. Here, in the practical sense, total values integrated over all wavelengths are shown. This approach is common in engineering. The two important rules are 1) that the sum of the three thermal radiation properties is unity; and 2) that absorptivity is equal to emissivity. Therefore, in this case, transmissivity = 1.0 – reflectivity – absorptivity.

Table. 2.3. Thermal radiation characteristics of covering materials
(after Okada in JGHA, 1987).

Films	Thickness, mm	Absorptivity	Transmissivity	Reflectivity
PE	0.05	0.05	0.85	0.1
	0.1	0.15	0.75	0.1
EVA	0.05	0.15	0.75	0.1
	0.1	0.35	0.55	0.1
PVC	0.05	0.45	0.45	0.1
	0.1	0.65	0.25	0.1
PETP	0.05	0.6	0.3	0.1
	0.1	0.8	0.1	0.1
	0.175	> 0.85	< 0.05	0.1
Non-woven		0.9	-	0.1
EVOH		> 0.9	-	< 0.1
Glass	3.0	0.95	-	0.05
PMMA		0.9	-	0.1
Aluminum powder mixed PE		0.65-0.75	-	0.25-0.35
Aluminized PE		0.15-0.4	-	0.6-0.85

PROBLEMS

1. What is the difference between mulch and floating mulch?

2. Spell out PE, PVC, EVA, FRP and FRA.

3. Describe the main differences in the thermal radiation characteristics of the covering materials PE, PVC and glass.

CHAPTER 3

DIGITAL SIMULATION

3.1. INTRODUCTION

Simulation of continuous systems started with the use of analog computers. Analog computers had been used widely, but they had several disadvantages. Time and magnitude scaling were cumbersome. Inaccuracies were caused by analog systems, which cannot separate signal change from noise in principle. There were also frequent breakdowns in hardware components. With the development of digital computers in the late 1960's and the growth in capacity of mainframes, digital simulation became the predominant technique for continuous simulation. In the 1980's, digital simulation spread to mini- and microcomputers. The high speed and large memory size of microcomputers have enabled us to use almost the same simulation languages that once were only available on mainframes.

The concept of digital simulation is the same with that used in analog computers. It is suitable for student lab work since PCs are now available in microcomputer labs in many schools.

3.2. SYSTEM DYNAMICS

The approach to simulating continuous systems, which is called System Dynamics has been well-known since it was used to simulate what would happen on the earth. The results of the simulation were published as "World Dynamics" by J. W. Forrester (1971) and later as "The Limits to Growth" by his successors such as D. H. Meadows (1972). The approach itself was developed originally by Forrester, and the simulation language **DYNAMO** was invented by him. It can be said that **DYNAMO** opened the era of digital simulation. The "World Dynamics" model of population expansion, energy resource expenditure, and pollution increase on the earth was so popular that its rather pessimistic predictions astonished public administrators in many countries in the 1970's. This kind of model cannot predict an event such as an energy crisis, which is caused politically and is discontinuous; when a crisis occurred, the rate of consumption of energy resources decreased, and the whole situation was then reconsidered and modified.

The language **DYNAMO** is more or less in the form of difference equations and has not been well used because more friendly languages such as **CSMP** were developed soon after. System dynamics is defined as a methodology to analyze system behaviour including feedback loops. The main applications are for analyses of social, biological, and ecological processes with many nested feedback loops and of non-linear systems.

3.3. SIMULATION LANGUAGES

Languages are classified into two groups: those for continuous systems and those for discrete systems. Languages such as **GPSS** and **SIMAN** are mainly for discrete systems; recent versions can handle continuous simulation too (Pegden, 1986), although all programs for continuous systems are more or less of the nature of **FORTRAN** subroutines. On the other hand, a continuous system simulation language such as **ACSL** can handle discrete simulation. Therefore, the two groups are getting closer, although there are still large discrepancies between the two. We are mostly interested in simulation of continuous systems, and it will be emphasized here.

Digital simulation languages have been developed to capitalize on the recent developments in simulation of continuous systems and the advantages of digital computers. They free us from the disadvantages of both time and magnitude scaling which were needed on analog computers. They take advantage of the higher accuracy of each component more than analog computers, and place no actual limitation on the number of functions used. Popular languages were **DYNAMO**, **MIMIC**, **DSL/90**, **CSMP** and **CSSL**.

The languages were then separated into two groups after transient languages such as **MIMIC** were discounted. One was the group represented by **CSMP** (Continuous System Modeling Program), developed by IBM, and the other was **CSSL**, which was mostly popular in Europe. The basic types of languages have similar capabilities, but with some differences in expressions. In biological research, it can be said that **CSMP** was still more popular than the other languages available on mainframes. PC versions of this kind of simulation language have been developed: for example, **micro-CSMP** (**CSMP** version for PCs and compatibles), **PCSMP** (similar to **micro-CSMP** with some hardware restriction), **ACSL** (Advanced Continuous System Language, an advanced version of **CSSL**) has much more flexibility but needs more typing in debugging processes and has hardware protection), and **SYSL** (80 - 90% compatibility with **CSMP**). **Micro-CSMP** was chosen as the language in the first edition of this book.

Since then, several new languages have been developed, such as **SIMNON**, **DYMOLA** and **Stella**. **SIMNON** is very similar to **ACSL** and **CSMP**. On the other hand, **DYMOLA** and **Stella** are called modelling languages and are more or less object-oriented languages (see Cellier, 1991). They have functions to make a model and can cooperate with other simulation languages such as **ACSL**, **SIMNON**, **MATLAB** (**SIMULINK**) and **FORTRAN**. **Stella** and **DYMOLA** introduce a relatively large number of functions for particular purposes and their functions might not be needed for the models in this book. Inputs and outputs are connected with special symbols. Flow diagrams are the main part of the program instead of description by equations. An ecological modelling group in the Netherlands that first developed **PCSMP** also developed **FST**(Fortran Simulation Translator). **FST** is more or less the same as **PCSMP** but is much more restricted by the tight grammar of **FORTRAN**, such as no mixture of integer and real numbers, than **PCSMP**. For example, **TO** can not be a variable since it is a part of the **GO TO** statement of **FORTRAN**.

A group of mathematical software has been developed including **Mathematica, MATLAB, Mathcad**, and **Maple**. These types of the software were originally developed to find formulas and perform mathematical calculations such as on matrices. **MATLAB** is linked with **SIMULINK**, which is a symbolic solver similar to analog computer techniques and **Stella** (see Bennett, 1995).

Since the operating system of PCs has changed from DOS to Windows and a Windows version of **micro-CSMP** or **PCSMP** does not exist, a new Windows-based simulation language is required to accompany the models developed in this book. **MATLAB (SIMULINK)**, product of MathWorks Inc., USA, is one of the most successful software packages currently available and is widely used for mathematical calculations. It is suited for work in control and in simulation as well. It is a powerful, comprehensive and user-friendly software package for performing mathematical computations. Equally as impressive are its plotting capabilities for displaying information. In addition to the core package, referred to as **MATLAB**, there are 'toolboxes' for several application areas. However, only the core package is required to solve the models of this book with one example in **SIMULINK**, one of the toolboxes.

The concept of any of these languages is based on the analog computer; that is the same. Once you become familiar with one simulation language, it is much easier to understand one of the others than to go from **FORTRAN** to **BASIC**.

It is clear that language is a tool, and you can make a model in a common language such as **FORTRAN** or **BASIC**. However, it is much easier to make a model in a simulation language and to understand a model someone else has developed than in a programming language, because a model in a simulation language is more like a set of mathematical equations than a list of programs.

3.4. DIGITAL SIMULATION BY CSMP AND MATLAB

3.4.1. Concept of the analog computer

The concept of analog computers is alive in digital simulation. One of the most important functions on analog computers is integration. Its symbol, shown in Fig. 3.1, is well-known to all engineering students. This figure exactly corresponds to the hardware configuration on analog computers. If dY/dt (derivative of an arbitrary function y due to time t is supplied into the operational amplifier whose function is integration, the original function Y can be obtained through an integrator which is shown by the integral sign. The minus sign is due to the hardware configuration. If input and output wires of the amplifier are connected together after the sign is converted, **exp(t)** is obtained as the output y. On analog computers you need not solve the equation. It can be said that the computer solves the differential equation you supply. In **SIMULINK**, the integrator is simply replaced by the function named 1/S, and without an inverter for sign change the same output is obtained.

Using digital simulation languages such as **CSMP**, the process is similar; that is, for example, in **CSMP** a function to integrate according to time (**INTGRL**) is available with other functions.

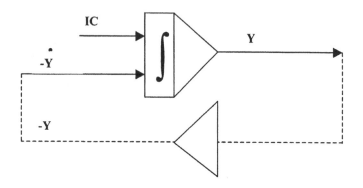

Figure 3.1. Diagram of integration on analog computers.

3.4.2. Comparison of CSMP and MATLAB programs with mathematical equations

Simulation of either a linear or a non-linear system, which is time-dependent -- in other words, dynamic and continuous -- can be done using analog computers. Simulation using analog computers has the advantage of being equation-oriented. Analog computers have many functions, such as integration, arbitrary function generation, and implicit expression.

Plant growth is a good example of non-linear and non-steady-state systems. In general, plant growth that is continuously changing can be expressed as a system of differential equations,

$$d\mathbf{Y1}(t)/dt = \mathbf{f1} \; (\; \mathbf{Y1}(t),\, \mathbf{Y2}(t),\, \ldots,\, \mathbf{Yn}(t),\, \mathbf{E1}(t),\, \mathbf{E2}(t),\, \ldots,\, \mathbf{Em}(t)\;)$$
$$d\mathbf{Y2}(t)/dt = \mathbf{f2} \; (\; \mathbf{Y1}(t),\, \mathbf{Y2}(t),\, \ldots,\, \mathbf{Yn}(t),\, \mathbf{E1}(t),\, \mathbf{E2}(t),\, \ldots,\, \mathbf{Em}(t)\;)$$

$$d\mathbf{Yn}(t)/dt = \mathbf{fn} \; (\; \mathbf{Y1}(t),\, \mathbf{Y2}(t),\, \ldots,\, \mathbf{Yn}(t),\, \mathbf{E1}(t),\, \mathbf{E2}(t),\, \ldots,\, \mathbf{Em}(t)\;)$$

where **Y1** to **Yn** are **n** state variables such as photosynthesis, and leaf and root weights, **E1** to **Em** are **m** environmental or boundary conditions such as solar radiation, temperature and CO_2 concentration, and **f1** to **fn** are functional relations which describe the rate of change of each state variable. With **n** unknown variables **Y1** to **Yn**, and **n** differential equations, unique solutions can be derived numerically.

These equations are expressed in an integral form in **CSMP** as

$$\mathbf{Y1(T)=INTGRL(IY1,\, f1(Y1(T),Y2(T),..,Yn(T),E1(T),E2(T),...,Em(T)))}$$
$$\mathbf{Y2(T)=INTGRL(IY2,\, f2(Y1(T),Y2(T),..,Yn(T),E1(T),E2(T),...,Em(T)))}$$

$$\mathbf{Yn(T)=INTGRL(IYn,fn(Y1(T),Y2(T),..,Yn(T),E1(T),E2(T), ... ,Em(T)))}$$

where **IY1, IY2, . . . , IYn** are initial conditions for **Y1, Y2, . . . , Yn**, respectively, and we still have **n** equations for **n** unknown variables.

Let's consider more details using a concrete expression. Suppose we have a mathematical expression as follows:

$d^2x/dt^2 = \mathbf{F} - \mathbf{A} * dx/dt - \mathbf{B} * \mathbf{x}$

where initial conditions are $\mathbf{x}(0) = \mathbf{X0}$, $d\mathbf{x}(0)/dt = \mathbf{DX0}$. Then a complete CSMP program for this is,

```
X = INTGRL (X0, DX)
DX = INTGRL (DX0, F - A * DX - B * X)
TIMER  FINTIM = 10.0, OUTDEL = 0.5, DELT = 0.1
PRTPLOT X
END
STOP
```

The main point is that the mathematical expression for plant growth is in the derivative form and the corresponding expression in **CSMP** is in integral form. The first two equations in the **CSMP** program are essential ones. It is not difficult to understand the correspondence between these two equations and the differential equation in the mathematical expression, if you notice that the **CSMP** program has the same variable **DX** (dx/dt in the mathematical expression) in the first two equations and that the first terms in the parentheses of the function **INTGRL**(INTeGRaL) are initial conditions.

The mathematical expression can be easily divided into two differential equations:

$$\mathbf{x} = \int dx/dt \, (dt)$$
$$dx/dt = \int (\mathbf{F} - \mathbf{A} * dx/dt - \mathbf{B} * \mathbf{x}) \, (dt)$$

TIMER in the **CSMP** program means time sets for numerical integration. **FINTIM** is the finish time, **OUTDEL** is the time step for the output and **DELT** is the time increment for integration. **PRTPLOT** (PrinT & PLOT) is a kind of output control, which gives a printout of numerical values as well as a printer plot. **TIME** is reserved as a system variable and is incremented by **DELT**. In **CSMP**, capital letters can only be used for expression, except in comments that start with an asterisk.

In **MATLAB**, two program files need to be generated to solve the above problem: A main program and an ODE function subprogram. A semicolon is required at the end of each command except for comments that start with a percent sign. As shown in Fig. 3.2, the dimensions of both **y0** and **dy** are 2 rows by 1 column. The dimensions of **y0** and **dy** need to be consistent.

```
% main.m
    T0=0; Tfinal=10;
    y0=[X0; DX0];
    [t,y]=ode23('subprg',[T0 Tfinal],y0);
    plot(t,y);
% subprg.m
    function dy=subprg(t,y)
    X=y(1);                              % define y1
    DX=y(2);                             % define y2
    DIFF_X = DX;                         % expression 1:  dy1/dt
    DIFF_DX=F - A * DX - B * X;          % expression 2:  dy2/dt
    dy=[ DIFF _X; DIFF _DX];             % dy=[expression 1; expression 2]
```

Figure 3.2. Structure of MATLAB programs for solving the differential equations.

The biggest advantage in **CSMP** or **MATLAB** programs is that the original equation need not be split into the form the computer can understand, as is always required in other common languages. Therefore, the program itself becomes very easily understandable for not only the programmer but also other people.

3.5. MODEL STRUCTURE AND REPRESENTATION

Models are classified into several categories. As we are concentrated into biological and environmental models, classification based on model structures should be noted. In order to understand how to make a model, the physical structure is the most important aspect to be considered. Thus, three main categories, lumped or distributed models, steady state or dynamic models and linear or non-linear models, are of primary concern. The **CSMP** and **MATLAB** languages are particularly powerful for solving dynamic non-linear system behaviour. To simplify the model, lumped models in which one variable is assigned to each object to express an average are very often used. For example, normally one variable is assigned to the inside air temperature of a greenhouse. However, if there is a large temperature gradient in one object, such as the soil layer, more than two variables can be assigned to the object even if the model is one-dimensional. The soil layer is divided into several layers, and in each soil layer temperatures are defined separately. This kind of model is called a distributed model. Air temperature in the greenhouse can also be divided into separate regions (see section 6.5).

3.5.1. Differential equation

Exponential growth can be expected if the relative growth rate of a living creature is constant. Let us consider population increase for humans. Assuming constant birth rate and no mortality, population increase rate is expressed as the product of the present population and the relative growth rate, that is,

$$d\mathbf{P}/d\mathbf{t} = \mathbf{BR} * \mathbf{P} \qquad (3.1)$$

where **P** is the present population and **BR** is the birth rate (births/unit time).

If **BR** is constant, eq. 3.1 is linear and is easily solved analytically. The solution is **P = A*exp (BR * t)**, assuming the initial condition of **P** is **A**. This is programmed in the following manner by **CSMP**:

$$\mathbf{P} = \mathbf{INTGRL}(\mathbf{A}, \ \mathbf{BR} * \mathbf{P}) \qquad (3.2)$$

where **INTGRL** is one of the powerful functions of **CSMP** used to integrate the second argument in the parentheses in terms of time. The initial condition is placed as the first argument in the same parentheses. The expression can be left as implicit. The solution of eq. 3.1 is programmed in the following manner in **MATLAB**:

```
% main.m
[t,y]=ode23('F', TSPAN, A)
% F.m
function  dy=F(t, y)
DIFF_P=BR * y;           % define y1=y , expression 1: dy1/dt
dy=[DIFF_P];
```

where **TSPAN** = **[T0 TFINAL]** integrates the system of differential equations **y'** = **F(t,y)** from time **T0** to **TFINAL** with initial condition **A**. **'F'** is the name of an **ODE** file.

Numerical integration is carried out in the background of this expression, and several integration methods are available. Therefore, this one expression is equivalent to approximately 20 to 50 **FORTRAN** statements for numerical integration. Since all calculations are conducted numerically, the system to be simulated is not necessarily linear.

If we consider death by disease or other reasons, eq. 3.1 can be changed to

$$d\mathbf{P}/d\mathbf{t} = \mathbf{P} * (\mathbf{BR} - \mathbf{DR}) \qquad (3.3)$$

where **DR** is mortality. This equation means that a small population increase ($d\mathbf{P}$) in a small time increment ($d\mathbf{t}$) is equal to net increase, the difference between population inflow (**BR * P**) and outflow (**DR * P**).

Sometimes flow charts or diagrams which were originally used for **DYNAMO** are used for **CSMP** and **MATLAB**, as shown in Fig. 3.3. This figure is the diagram of eq. 3.3. The population level is express by a rectangular shape. **BR** and **DR** are flow rates, and straight lines show the flow of people. In this figure, it is assumed that people flow from a kind of source to a kind of sink, both of which are indicated with cloud-like symbols. Valves can change flow rates in an actual flow, such as water in pipelines.

The shape of the valve is used to calculate flow rates, **BR** and **DR** in the present case. This kind of flow diagram can be an aid to understanding the program, but sometimes it is cumbersome, while direct translation from mathematical equations is straight- forward. In the present textbook, therefore, no flow charts or diagrams are used.

This concept of the relationship between flow and level can be applied to similar systems such as heat flow and mass flow in the air and in soil.

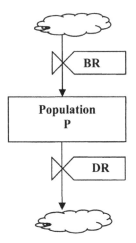

Figure 3.3. Flow diagram of eq. 3.3.

3.5.2. Description of systems

The important point in the description of systems is that the basic concept of the model is a flow of something: for example, it can be heat, water vapor or carbon dioxide in the air. Therefore, the governing rule is conservation of these components.

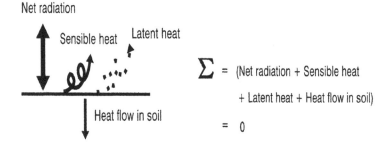

Figure 3.4. Energy balance equation.

In Fig. 3.4, energy balance of the soil surface is considered. We are assuming a hypothetical very thin film soil layer at the surface, which does not have appreciable volume to store energy in it. Since energy conservation holds here, the summation of all energy inflow and outflow is zero. In other words, net incoming energy is equal to net outgoing energy. This is the basic concept of how to build up an equation to describe a system. Fig. 3.5 shows how to build a differential equation. Let's assume there is a mass of air or water of which the volume is V, density is r, and volumetric heat capacity is C_p. Energy Q_1 and Q_3 are coming in and Q_2 is going out from the mass. Then, the energy change of this mass in dt time is expressed as dQ/dt, and it is clearly equal to $Q_1 - Q_2 + Q_3$ (the difference of inflow and outflow). This energy change can be easily converted to temperature change by introducing thermal properties of the mass as shown in the figure. This is a basic differential equation to show the temperature change of the mass.

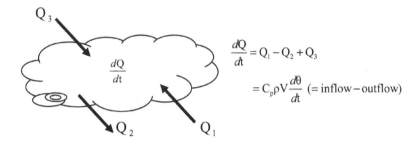

$$\frac{dQ}{dt} = Q_1 - Q_2 + Q_3$$

$$= C_p \rho V \frac{d\theta}{dt} \quad (= \text{inflow} - \text{outflow})$$

Figure 3.5. Differential equation to express an energy balance.

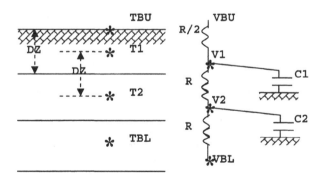

Figure 3.6. Heat flow and temperature regime in the soil layer:
(a). Left: Heat flow, (b). Right: Electric network analogy.

3.5.3. Heat flow and temperature regime in the soil

Heat flow in the soil is complicated because heat flow is associated with water flow. However, in most cases, it is sufficient to consider heat flow using apparent thermal conductivity, which includes the effect of water flow. Then heat flow in the soil is that in a solid body.

Referring to Fig. 3.6a, let us suppose the flat ground is divided into three even layers in depth and the temperatures at the middle of each layer are **T1**, **T2** and **TBL**, respectively. The bottom temperature **TBL** is a boundary condition and the temperature at the surface of the ground is **TBU**.

Referring to this analogy and the scheme in Fig. 3.5, the following equations are derived, as the temperature increase in the layer in a small time increment is equal to the total heat flow in the layer considered from the surroundings, in this case the upper and lower layers:

It is clear from the same figure that an electric passive network system, which is called π (pi) network, can represent the heat flow in the soil layer (see Fig. 3.6b).

$$CS*VS*(dT1/dt) = KS*AS*((TBU-T1)/DZ*2+(T2-T1)/DZ) \qquad (3.4)$$

$$CS*VS*(dT2/dt) = KS*AS*((T1-T2)/DZ+(TBL-T2)/DZ) \qquad (3.5)$$

where **CS** is volumetric heat capacity of the soil, **VS** is the volume of the soil layer, **AS** is the surface area of the soil layer, **DZ** is the thickness of one layer, **KS** is the thermal conductivity of the soil, and **t** is time. We assumed that heat flows in from the upper and the lower layers. This is appropriate if we consider properly the signs involved. If the flow is opposite previously assumed direction, then the sign is inverted automatically.

These two equations are in the differential form. Again, they can be rearranged into the integral forms of **CSMP** (eqs. 3.6a and 3.7a) and the differential forms of **MATLAB** (eqs. 3.6b and 3.7b) as shown below:

$$T1 = INTGRL(IT1, KS*((TBU - T1)*2.0 + (T2 - T1))/ DZ/DZ/CS) \quad (3.6a)$$

$$T2 = INTGRL(IT2, KS*((T1 - T2) + (TBL - T2))/ DZ/DZ/CS) \qquad (3.7a)$$

$$DIFF\_T1 = KS*((TBU - T1)*2.0 + (T2 - T1))/ DZ/DZ/CS \qquad (3.6b)$$

$$DIFF\_T2 = KS*((T1 - T2) + (TBL - T2))/ DZ/DZ/CS \qquad (3.7b)$$

With two unknowns, **T1** and **T2**, and two boundary conditions, **TBU** and **TBL**, the system can be solved.

Let's assume that heat flow in the soil layer shown in Fig. 3.7 is two-dimensional, with vertical and horizontal flows. Each heat flow is proportional to the temperature gradient and thermal conductivity and inversely proportional to the distance.

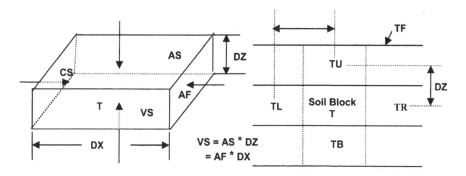

Figure 3.7. Basic diagram of heat flow in a system.

Suppose each temperature shown in the figure is at the center of a soil block. Then the differential equation for the system is:

$$\mathbf{CS * VS} * (d\mathbf{T}/d\mathbf{t}) = \mathbf{KS} * (\mathbf{AS} * ((\mathbf{TU\text{-}T}) / \mathbf{DZ} + (\mathbf{TB\text{-}T}) / \mathbf{DZ}) +$$
$$\mathbf{AF} * ((\mathbf{TR\text{-}T}) / \mathbf{DX} + (\mathbf{TL\text{-}T}) / \mathbf{DX})) \qquad (3.8)$$

where **KS** is the thermal conductivity of the soil, **AS** is soil surface area for vertical heat flow, **AF** is that for horizontal heat flow, **DX** is the horizontal distance from center to center of soil blocks, **DZ** is the vertical distance from center to center, and **CS** is the heat capacity of the center soil block.

The differential form is more common in mathematical expressions, but the integral form is straightforward for programming in simulation languages. Therefore, the **CSMP** and **MATLAB** expressions for eq. 3.8 are as follows:

$$\mathbf{T = INTGRL(IT, KS} * (((\mathbf{TU - T}) + (\mathbf{TB - T})) / \mathbf{DZ} / \mathbf{DZ} \ldots$$
$$+ ((\mathbf{TR - T}) + (\mathbf{TL - T})) / \mathbf{DX} / \mathbf{DX}) / \mathbf{CS}) \qquad (3.9a)$$

$$\mathbf{DIFF\_T = KS} * (((\mathbf{TU - T}) + (\mathbf{TB - T})) / \mathbf{DZ} / \mathbf{DZ} + \ldots$$
$$((\mathbf{TR - T}) + (\mathbf{TL - T})) / \mathbf{DX} / \mathbf{DX}) / \mathbf{CS} \qquad (3.9b)$$

Model representation is not necessarily explicit. The form of integration always includes implicit expression, and the function **IMPL** in **CSMP** or a similar function **fzero** in **MATLAB** can be used to solve implicit functions except for integrals. Programming is straightforward and can be a kind of translation of mathematical equations into the expressions in each language. Therefore, the following several examples are the best way to understand programming and thus modelling. An ellipsis (...) indicates that the statement continues to the following line in the **CSMP** and **MATLAB** programs.

3.6. A MODEL FOR TEMPERATURE REGIMES IN THE SOIL (**CUC01**)

3.6.1. *Model description*

Fig. 3.8 shows the Command Window of **MATLAB** running the **cuc01** model. The user can enter '**cuc01**' or '**cuc01(n)**' to run the model, where n is 1 to 4. Entering '**cuc01**' will have the same result as with '**cuc01(1)**'. Fig. 3.9 shows the result of the CUC01 model in the Figure Window. Fig. 3.10 shows the scripts of the model, which consists of a main program (Fig. 3.10a) and a function subprogram (Fig. 3.10b). In the **MATLAB** program, **%** denotes that the line is a comment only. The command '**function cuc01(trial)**' is the first line of the script, in which '**cuc01**' is the name of the function and needs to be consistent with the file name. The variable 'trial' contains the number carried into the program. If there is no argument or the argument value is larger than 4 or smaller than 1, the value of 1 will be assigned to 'trial'. Based on the '**trial**' value, the '**switch...case...end**' provides branching with various pairs of **ks** and **cs** values. The thermal conductivity of the soil, **ks** indicates how fast the heat will transfer through the soil, and the heat capacity of the soil, **cs** indicates how much heat can be store by the soil. These values are declared as **global** variables in the second line of the scripts listed in Fig. 3.10a and Fig. 3.10b.

The matrix variable **y0** contains the initial temperatures of T_1 to T_5. As listed in Fig. 3.10a, **y0** is a 5 by 1 matrix (column vector). The **y0** matrix is then fed to the subprogram 'soil01.m' using the **ode23()** function. The core of the main program of the CUC01 model lies in the following command: **[t, y] = ode23('soil01',[tstart tfinal],y0);**

Several functions can be used in solving simultaneous ordinary differential equations, including **ode23, ode45, ode113, ode15s, ode23s, ode23t** and **ode23tb**. The major difference among these functions is in the numerical methods used. For non-stiff differential equations, **ode23, ode45** and **ode113** can be used. For stiff differential equations, **ode15s** or **ode23s** can be used. More details can be found from online help of **MATLAB**; for example, type '**help ode23**' in the Command Window followed by the Enter key to learn more about **ode23**. Some helpful information is listed in the last section of this Chapter.

The last line of the 'soil01.m' subprogram is the **dy** matrix. The dimensions of this matrix must be consistent with the **y0** matrix listed in '**cuc01.m**'.

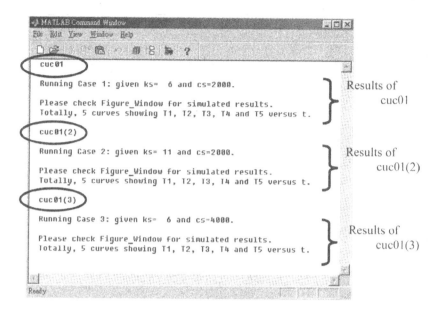

Figure 3.8. Command Window of MATLAB running cuc01 model.

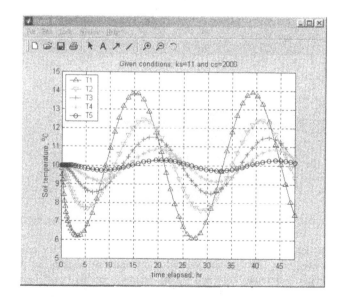

Figure 3.9. Figure Window of MATLAB running cuc01 model.

```
%       Temperature regime in the soil layer              CUC01.m
%       Boundary condition is surface temperature.
%       Function required: soil01.m
%
function cuc01(trial)
global ks cs
if nargin==0                    % If no argument
   trial=1;                     % use 1 as the default.
end
if trial>4 | trial <1           % If argument_value >4 or <1
   trial =1                     % use 1 as the default
end
switch trial
case 1
      ks=5.5;cs=2000;           % default values
case 2
      ks=11;cs=2000;            % ks doubled
case 3
      ks=5.5;cs=4000;           % cs doubled
case 4
      ks=11;cs=4000;            % ks, cs both doubled
end
% ks: Soil thermal conductivity (kJ/m/C) and ks/3.6  (W/m/C)
% cs: Heat capacity of soil  (kJ/m3/C)
%
tstart = 0;     tfinal = 48;
y0=[10;10;10;10;10];            % 5x1 matrix for initial conditions.
[t,y] = ode23t('soil01',[tstart tfinal],y0);
% calling function ode23t with constants and eqs. in 'soil01.m'
%                       with simulated time from tstart to tfinal
%                       with initial conditions in matrix y0 and
%                       with calculated answer in matrix y.
%
plot(t,y(:,1),'b^-',t,y(:,2),'gV-',t,y(:,3),'r+-',…
               t,y(:,4),'c*-',t,y(:,5),'ko-');
axis([-inf,inf,5,15]);
grid on;
xlabel('time elapsed, hr');
ylabel('Soil temperature, ^oC');
tit=['Given conditions: ks=',num2str(ks),' and cs=', num2str(cs)];
title(tit);
legend('T1','T2','T3','T4','T5',2);
fprintf('\n  Running Case %1.0f: given ks=%3.0f and cs=%4.0f. \n\n',…
        trial,ks, cs);
disp('  Please check Figure_Window for simulated results.');
disp('  Totally, 5 curves showing T1, T2, T3, T4 and T5 versus t.');
```

Figure 3.10a. Main program of model for temperature of soil layers (**CUC01.m**).

The command '**plot(t, y(:,1),'b^-',....)**' draw five curves on the Figure Window as shown in Fig. 3.9. The first and second arguments are the data for the x and y axes. The length of the matrices **t** and **y(:,m)**, where **m** equals 1 to 5, need to be the same. The third argument is a character string that defines line types, plot symbols and color to be displayed on the screen. Entering '**help plot**' in the Command Window can reveal all the possible combinations as listed below.

y	Yellow	.	Point	-	Solid
m	Magenta	O	Circle	:	Dotted
c	Cyan	X	x-mark	-.	Dashdot
r	Red	+	Plus	--	Dashed
g	Green	*	Star		
b	Blue	s	Square		
w	White	d	Diamond		
k	Black	v	Triangle (down)		
		^	Triangle (up)		
		<	Triangle (left)		
		>	Triangle (right)		
		p	Pentagram		
		h	Hexagram		

The command 'axis' allows self-arrangement on both x and y axes. The '-inf' and 'inf' stand for no preset lower bound (LB) and upper bound (UB) of this axis. The first two parameters are for the LB and UB of the x axis and the third and fourth parameters are for the LB and UB of the y axis, respectively.

There are four commands frequently used after the **plot** command. Commands 'xlabel('text')' and 'ylabel('text')' allow the user to assign text to the x and y axes; commands 'title('text')' and 'legend('text1','text2',...,pos)' allow the user to assign text to the title of the plot and to the legend, respectively. The last argument 'pos' of 'legend()' places the legend in the specified location:

0 = Automatic "best" placement (least conflict with data)
1 = Upper right-hand corner (default)
2 = Upper left-hand corner
3 = Lower left-hand corner
4 = Lower right-hand corner
-1 = To the right of the plot

The 'fprintf(format,a,...)' command writes formatted data to the screen, and 'format' is a string containing C language conversion specifications. Conversion specifications involve the character %, optional flags, optional width and precision fields, optional subtype specifier, and conversion characters d, i, o, u, x, X, f, e, E, g, G, c, and s. For more details, see the 'fprintf' function description in the online help or refer to a C language manual. The special formats \n,\r,\t,\b,\f can be used to produce linefeed, carriage return, tab, backspace, and form feed characters respectively. Use \\ to produce a backslash character and %% to produce the percent character. The command 'disp(x)' can display an array on the screen. If **x** is a string of text, the text is displayed. The results of the last three commands of the script listed in Fig. 3.10a can be found in Fig. 3.8 as indicated by the large '}' sign.

Fig. 3.10b lists the script of 'soil01.m'. Since the **ode** function must return a column vector, **dy** is the column vector to be returned as listed in the second line from the bottom of Fig. 3.10b. In the script for calculating **TF**, 'pi' is used, and is a reserve word of **MATLAB**, representing π. In the subprogram 'soil01.m', there are 5 unknowns, **y(1)** to **y(5)** with 5 ordinary differential equations.

```
% Subprogram to be used with cuc01.m                    soil01.m
function dy = soil01(t,y)
global ks cs
z=0.1; % Depth of each soil layer (m)
T0=10; % Average outside temperature (C)
TU=5;  % Amplitude, temperature variation
TBL=10;% Boundary soil temperature (C)
TF=T0+TU*sin(2*pi/24.*(t-8));
% t is time (in hours)
% TF: Soil temperature of surface layer (C)
%      Maximum temperature occurs
%      at 2 o'clock in the afternoon
T1=y(1);T2=y(2);T3=y(3);
T4=y(4);T5=y(5);
DIFF_T1=2*(TF-T1)+(T2-T1);
DIFF_T2=(T1-T2)+(T3-T2);
DIFF_T3=(T2-T3)+(T4-T3);
DIFF_T4=(T3-T4)+(T5-T4);
DIFF_T5=(T4-T5)+(TBL-T5)*2;
val=ks/z/z/cs;
dy = [DIFF_T1; DIFF_T2; DIFF_T3; DIFF_T4; DIFF_T5]*val;
% Format of dy should be consistent with y0 in cuc01.m (5x1 matrix)
```

Figure 3.10b. Subprogram of cuc01 model (**soil01.m**) *and soil diagram.*

3.7. APPLICATION TO STEADY STATE MODELS

Once the dynamic model for non-steady-state conditions has been developed, it can be easily applied to steady-state conditions. For example, if the two boundary conditions in eqs. 3.4 and 3.5 are constant, the left-hand side of the equations will become zero after a certain period of time. Then, the temperature gradient would be linear, and each temperature would be found by interpolation using the physical properties of the soil layer.

3.8. MORE ON MATLAB

Fig. 3.11 shows 'cuc01a.m', which is an expansion of 'cuc01.m' with more **MATLAB** commands included. The command 'tic' starts a stopwatch timer; 'toc' reads the stopwatch timer. The execution time required between commands 'tic' and 'toc' will be displayed on the screen upon the execution of 'toc'. Note that the **y0** matrix looks different from the one listed in 'cuc01.m'; however, they are in fact the same. The **y0** listed in Fig. 3.10a is a 5 by 1 matrix (column vector) and the **y0** listed in Fig. 3.11 is the transpose matrix of a 1 by 5 matrix. The transpose matrix of a row vector is again, a column vector.

The command '**h1=findobj('tag','Temperature')**' will find the object using 'Temperature' as a tag name and assign the handle to the **h1** variable. Following by '**close(h1)**' will close the h1 object, that is the one with 'Temperature' as a tag name. Adding these two commands, prior to the '**figure()**' command can prevent opening too many Figure Windows with the 'Temperature' tag name if the program is executed several times.

The command '**figure()**', by itself, creates a new Figure Window. Many properties were involved in the Figure Window such as '**tag**', "**Resize**", '**MenuBar**', '**Name**', '**NumberTitle**', '**Position**', etc. Fig. 3.9 was created without using the '**figure()**' command and Fig. 3.12 was created with the '**figure()**' command listed in the '%--[Figure 1]--' section of the script in Fig. 3.11. Fig. 3.12 has a user-defined 'Name', that is the text written at the top of the Figure Window, and also is without the command menu and icons listed in the second and third rows from the top of Fig. 3.9.

Assigning **figure** to a handle using the command '**h=figure(...)**', followed by '**get(h)**', will generate a list of figure properties and their current values. More details can be found in online help.

The command '**h=plot(.....)**' assigns the plotting operation to a handle '**h**', allows future manipulation on this plot such as setting the line width, and returns its handle.

```
%     Temperature regime in the soil layer                CUC01a.m
%     Boundary condition is surface temperature
%     Function required: soil01.m
%
function cuc01a(trial)
global ks cs
if nargin==0 | trial>4 | trial <1,  trial =1;          end
switch trial
          case 1, ks=5.5;       cs=2000;
          case 2, ks=11;        cs=2000;       % ks doubled
          case 3, ks=5.5;       cs=4000;       % cs doubled
          case 4, ks=11;        cs=4000;       % ks, cs both doubled
end
%---[Core]-------------------------------------------------------
tic                                  % start recording time
tstart = 0;tfinal = 48;
IT1=10;IT2=10;IT3=10;IT4=10;IT5=10;
y0 = [10 10 10 10 10]' ;  % Transpose of row matrix is column matrix
[t,y] = ode23('soil01',[tstart tfinal],y0);
toc                        % show elapsed time from tic to toc
%---[Figure1]----------------------------------------------------
h1=findobj('tag','Temperature'); close(h1);
% prevent from opening too many same figure windows
figure('tag','Temperature','Resize','on','MenuBar','none',...
     'Name','CUC01a.m (Figure 1: Temperatures in 5 soil layers)',...
     'NumberTitle','off','Position',[160,80,520,420]);
h=plot(t,y(:,1),'k-*',t,y(:,2),'b:o',t,y(:,3),'r:^',t,y(:,4), ,...
     'go-',t,y(:,5));
set (h,'linewidth',2); axis([-inf,inf,5,15]);grid on;
xlabel('time elapsed, hr'); ylabel('Soil temperature, ^oC');
legend('T1','T2','T3','T4','T5');
```

```
%---[Figure2]--------------------------------------------------------
h2=findobj('tag','Temp5');close(h2);
h2=figure('tag','Temp5','Resize','on','MenuBar','none',...
    'Name','CUC01a.m (Figure 2: Temperature in each soil layer)',...
    'NumberTitle','off','Position',[200,40,520,420]);
figure(h2); subplot(5,1,1);    plot(t,y(:,1),'k-*'); ylabel('T1');
% draw the 1st plot out of 5 row x 1 col. plots per figure
axis([-inf,inf,5,15]); grid on;
subplot(5,1,2);         plot(t,y(:,2),'b:o'); ylabel('T2');
% draw the 2nd plot out of 5 row x 1 col. plots per figure
axis([-inf,inf,5,15]); grid on;
subplot(5,1,3);         plot(t,y(:,3),'r:^'); ylabel('T3');
axis([-inf,inf,5,15]); set(gca,'ytick',[8 10 12]);    grid on;
subplot(5,1,4);         plot(t,y(:,4),'go-'); ylabel('T4');
axis([-inf,inf,5,15]); set(gca,'ytick',[8 10 12]);    grid on;
subplot(5,1,5);         plot(t,y(:,5));       ylabel('T5');
xlabel('time elapsed, hr');
axis([-inf,inf,5,15]); set(gca,'ytick',[8 10 12]);    grid on;
clc;    % clear command window
disp('  Thank you for using CUC01a.');    disp(' ');
disp('  You can enter ''close all'' to close Figure_Windows.');
```

Figure 3.11. Main program of CUC01a model (**CUC01a.m**).

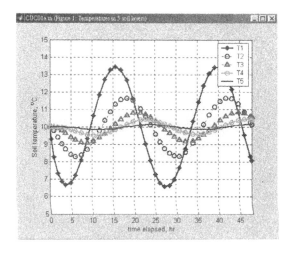

Figure 3.12. Figure generated from '%--[Figure 1]--' section of cuc01a.m.

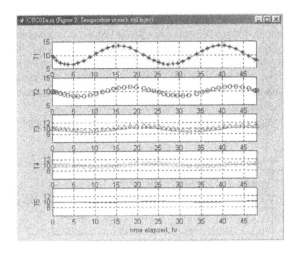

Figure 3.13. Figure generated from '%--[Figure 2]--' section of cuc01a.m.

The script listed in the '%--[Figure 2]--' section of Fig. 3.11 generates Fig. 3.13. The command **'figure(h)'** makes h the current figure, forces it to become visible, and brings it to the foreground, in front of all other figures on the screen. If Figure h does not exist and h is an integer, a new figure is created with handle **h**. The command **'subplot(m,n,p)'** breaks the Figure Window into an m-by-n matrix of small axes, and selects the p-th axes for the current plot. The command **'plot()'** following **'subplot(...,p)'** draws the plot in the p-th axes. As shown in Fig. 3.13, there are five subplots in one Figure Window. The last three subplots have different y ticks compared with the first two subplots. These y ticks of the last three subplots were generated using the **'set(gca,'ytick',[8 10 12])'** command. The term 'gca', representing 'get handle to current axis', is a reserve word in **MATLAB**. Both 'ytick' and 'xtick' can be assigned by the user with the **'set(gca,..)'** command. The command **'clc'** is used to clear the Command Window.

3.9. SIMULINK

SIMULINK is one of the toolboxes linked with **MATLAB** and is suitable for dynamic simulation. It is a kind of graphical approach and is based on the concept of analog computers as shown in Fig. 3.14. This figure shows the model CUC01 in **SIMULINK**. Fig. 3.1 can be a step to understanding this figure. In Fig. 3.14, the integrator is labeled 1/S, the summers are shown as squares with plus and minus signs, and the coefficients to be multiplied are in triangles. The construction of the model is straightforward. The flow is from left to right. There are two boundary conditions, the soil surface temperature change is given by a sine wave (all parameters are hidden under each symbol) plus a constant, and the bottom temperature is given as a constant 10. The scope symbol is the output, and any of the outputs T1 through T5 can be seen through the scope. It is apparent that the first line components are all for the temperature of the first soil layer, T1. T1 is the output of the first integrator and is fed back as an input to the summers with a minus sign. Then, the first boundary condition, the soil surface temperature minus T1 is one of the two inputs to the next summer. Following this approach, the whole diagram can be understood.

In this book, models in **SIMULINK** are not included because of the space limitations.

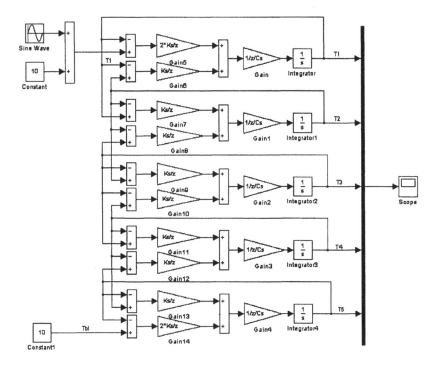

Figure 3.14. The model CUC01 in SIMULINK.

MATLAB FUNCTIONS USED

%	Comments.
;	Prohibit from display to the monitor.
:	Represent a complete row or column of a matrix.
...	Continue in next line.
axis	Control axis scaling and appearance. **Axis**([XMIN XMAX YMIN YMAX]) sets scaling for the x- and y-axes on the current plot.
clc	Clear command window.
disp	Display array. **Disp**(X) displays the array, without printing the array name. In all other ways, the same as leaving the semicolon off an expression except that empty arrays don't display. If X is a string, the text is displayed.
figure	Creates a new figure window, and returns its handle.
findobj	Find objects with specified property values.
fprintf	Write formatted data to screen.
global	Define global variable.
grid	Grid lines. **Grid on** adds grid lines to the current axis. **Grid off** takes them off. **Grid** , by itself, toggles the grid state of the current axis.
gca	Get handle with Current Axis.
legend	Graph legend. **Legend**(string1,string2,string3, ...) puts a legend on the current plot using the specified strings as labels. **Legend**(...,Pos) places the legend in the specified location:
	$\quad$ 0 = Automatic "best" placement (least conflict with data)
	$\quad$ 1 = Upper right-hand corner (default)
	$\quad$ 2 = Upper left-hand corner
	$\quad$ 3 = Lower left-hand corner
	$\quad$ 4 = Lower right-hand corner
	$\quad$ -1 = To the right of the plot
num2str	Convert number to string.
ode23	Solve non-stiff differential equations, low order method. (**MATLAB** 4 and higher versions) [T, Y] = **ode23**('F',TSPAN,Y0) with TSPAN = [T0 TFINAL] integrates the system of differential equations y' = F(T,Y) from time T0 to TFINAL with initial conditions Y0. 'F' is a string containing the name of an ODE file. <u>Function F(T, Y) must return a column vector.</u> Each row in solution array Y corresponds to a time returned in column vector T.
ode45	Solve non-stiff differential equations, medium order method. (**MATLAB** 4 and higher versions)
ode113	Solve non-stiff differential equations, variable order method. (**MATLAB** 5 and higher versions)
ode15s	Solve stiff differential equations and DAEs, variable order method. (**MATLAB** 5.2 and higher versions)
ode23s	Solve stiff differential equations, low order method.

	(**MATLAB** 5.2 and higher versions)
ode23t	Solve moderately stiff ODEs and DAEs, trapezoidal rule.
	(**MATLAB** 5.2 and higher versions)
ode23tb	Solve stiff differential equations, low order method.
	(MATLAB 5.2 and higher versions)
Plot	Linear plot. **Plot**(X,Y) plots vector Y versus vector X.
Set	Set object properties. **Set**(H,'PropertyName',PropertyValue) sets the value of the specified property for the graphics object with handle H. H can be a vector of handles, in which case SET sets the properties' values for all the objects.
subplot	Create axes in tiled positions.
Tic	Start a stopwatch timer.
Title	Graph title. **Title**('text') adds text at the top of the current axis.
Toc	Read a stopwatch timer.
xlabel	X-axis label. **Xlabel**('text') adds text beside the X-axis on the current axes.
Ylabel	Y-axis label. **Ylabel**('text') adds text beside the Y-axis on the current axes.

PROBLEMS

1. Develop the energy balance equations.

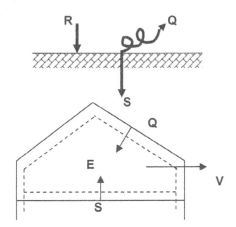

   a) On the bare ground surface, assume
      absorbed radiation is R (J/s/m^2),
      out-going heat flux mostly by
      convection is Q (J/s/m^2), and heat
      flux into the soil is S (J/s/m^2).

   b) Heat (E) is stored in the air mass
      in the greenhouse after
      obtaining heat (Q) from the
      covering surface, S from the
      ground surface, and V by
      ventilation. Units are all (J/s).

   c) For a single horizontal leaf of area
      A (cm^2), assume no heat
      capacity of the leaf and solar
      radiation absorbed is S (W/cm^2),
      net long wave radiation
      (effective radiation) on the upper
      side is R_u and that at the lower
      side is R_l (J/cm^2/s), heat
      convection from the leaf is Q
      (J/cm^2/s), transpiration is q
      (g/cm^2/s), and latent heat of
      vaporization is L (J/g).

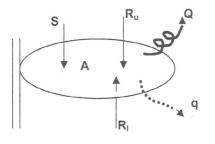

2. Develop the differential
   equations.

   a) Direct solar radiation R
      penetrated into a plant
      canopy follows Lambert-
      Beer's law. Assuming
      radiation just above the
      canopy is R_0 and extinction
      coefficient in the canopy is
      k, describe the penetrated
      radiation rate in terms of the
      depth from the top of the canopy.

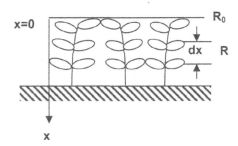

b) A coffee cup is filled with water is
heated by Q (J/s), and the over-all heat
loss from the cup is L (J/s). Describe
temperature increase in the cup. Assume
the amount of water in the cup is W (g),
the heat capacity of water is C_p
$(J/°C/cm^3)$ and its density is 1 (g/cm^3).

c) Describe the carbon dioxide
concentration change in the
greenhouse, assuming
outside concentration is C_o
(ppm) and constant, inside is
C (ppm), no generation from
soil, consumption by plant
photosynthesis is P
$(mg/cm^2/min)$, total leaf area
is A (cm^2), greenhouse

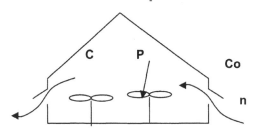

volume is V (m^3), and ventilation rate is n (1/h). Note that 1 mole of carbon
dioxide (44 g) is equivalent to 22400 cm^3 at 0 °C and 1 atm. More discussion
on the concentration units, $\mu l/l$, ppm, vpm, and $\mu mol/mol$ is given in
section 10.5.

3. Write the following differential equations in **MATLAB**.

a) $dy/dt = - A * y$ and $y_{t=0} = B$
b) $dy/dt = \cos(y)$ and $y_{t=0} = 1$
c) $d^2x/dt^2 = F - A * dx/dt - B * x$ and $x_{t=0} = X0$; $dx/dt_{t=0} = DX0$
d) Host-parasite or predator-prey model
$dH/dt = (K_1 - K_2 * P) * H$ and $H_{t=0} = H0$
$dP/dt = (- K_3 + K_4 * H) * P$ and $P_{t=0} = P0$
e) Shells and limpets model
$dS/dt = K_1 * S - K_2 * S^2 - K_3 * L$ and $S_{t=0} = S0$
$dL/dt = B * K_3 * S * L - K_4 * L - K_5 * L$ and $L_{t=0} = L0$

4. Develop **MATLAB** programs to calculate the following equations and run from
time t = 0 to 5.

a) Growth curve
$y = \exp(t)$
$dy/dt = y$ and $y_{t=0} = 1$
b) Decay curve
$y = 10 * \exp(- 0.1 * t)$
$dy/dt = - 0.1 * y$ and $y_{t=0} = 10$

c) Periodic curve

$y = 3* \sin (0.6 * t)$

$d^2y/dt^2 = - 0.36 * y$ and $dy/dt_{t=0} = 1.8 ; y_{t=0} = 0$

d) Response curve

$y = 1 - \exp (- 3 * t)$

$dy/dt = 3 - 3 * y$ and $y_{t=0} = 0$

e) Rectangular hyperbola (Michaelis-Menten relation)

$u = k * t / (K + t)$

$du/dt = k*K/(K + t)^2$ and $u_{t=0} = 0$

f) Logistic curve

$W = W_i * W_f * \exp (W_f * k * t) /(W_f - W_i + W_i* \exp (W_f * k * t))$

$dW/dt = k* (W_f - W)* W$ and $W_{t=0} = W_i$

5. Derive that $dX1/dt = A * X1 - X2$ and $dX2/dt = - X1$ are equivalent to $d^2X1/dt^2 = A * dX1/dt + X1$.

6. Write a **MATLAB** statement equivalent to eq. 3.3, assuming the initial condition of **P** is **A**.

7. Modify the program **CUC01**, assuming the thermal conductivity of soil **KS** is a function of temperature. Use the expression **KS = 5.5 + 0.1 * TEMP**, where **TEMP** is soil temperature.

8. Derive the system of differential equations from Figure 3.14. Note: The input sine function is given in Figure 3.10b.

CHAPTER 4

HEAT BALANCE OF BARE GROUND

4.1. INTRODUCTION

The basic component in the system we are considering is bare ground. In the daytime, its surface is heated by solar radiation when the weather is fine. The surface also loses heat to the cold sky by long wave radiation. It can be visualized that heat is transferred by conduction to the lower soil layers, neglecting water movement in the soil. At the surface, not only sensible but also latent heat transfer occurs. These heat flows from the surface to the ambient air are by convection. Therefore, three types of heat transfer are at work in a soil-air system, as shown in Fig. 4.1.

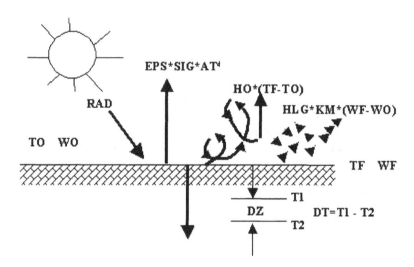

Figure 4.1. Heat balance of bare ground

Heat conduction is heat flow due to molecular movement and is predominant in solid bodies, in which the other two types of heat transfer do not occur. The amount of heat per unit time per unit area is proportional to the product of the thermal conductivity of the material and the temperature difference, and inversely proportional to the distance where the temperature difference occurs:

$$Q = -KS * DT / DZ \qquad (4.1)$$

where **Q** is heat flow (W/m^2), **KS** is thermal conductivity (W/m/$^{\circ}$C), **DT** is temperature difference ($^{\circ}$C), and **DZ** is a distance (m) short enough that we can neglect heat stored in this thin layer.   The negative sign indicates that the direction of heat flow is opposite to the temperature gradient.

Heat transfer due to convection is given by

$$Q = H * DT = H * (TF - TO) \qquad (4.2)$$

where **Q** is heat flow (W/m^2), **H** is the heat transfer coefficient due to convection (W/m^2/$^{\circ}$C), and **DT** is the temperature difference between the surface **TF** and the ambient fluid, in this case the air **TO** ($^{\circ}$C).

Latent heat transfer is based on vapor flow due to the vapor gradient between the surface and the air.   The vapor potential is expressed as the actual content of vapor in the air on either a weight or a volume basis.   Here, we use the weight basis. Therefore, the vapor content of a unit weight of air is expressed as **W** (kg/kg DA) vapor weight per unit weight of dry air (which does not include vapor), where kg DA means unit weight of dry air.   Dry air basis is used because it is not affected by the amount of vapor involved and does not vary.   Wet air, therefore, is the total of the dry air and vapor.   The equation to express the latent heat flow is then,

$$Q = HLG * KM * DW / 3.6 = HLG * KM * (WF - WO) / 3.6 \qquad (4.3)$$

where **Q** is heat flow (W/m^2), **HLG** is latent heat due to vaporization (2501.0 kJ/kg), **KM** is mass transfer coefficient (kg/m^2/hr), and **DW** is the difference of humidity ratio between the surface (**WF**) and the air (**WO**) (kg/kg DA).   At the soil surface, it is assumed that the air is usually saturated with vapor and that the actual exchange takes place very close to the surface.   When the soil surface is dry, a wetness factor coefficient is introduced to express how much **WF** differs from the saturation value.

Radiation heat transfer occurs without any transferring medium such as the air. This means that heat transfers directly from one surface to the other through the air. The ruling relationship of this transfer is given as:

$$Q = EPS * SIG * AT^4 \qquad (4.4)$$

where **Q** is heat flow (W/m^2), **EPS** is a proportional constant between 0 to 1 called emissivity (ND) which is dependent on the material, **SIG** is the Stefan-Boltzmann constant and is 5.67 x 10^{-8} (W/m^2/K^4), and **AT** is absolute temperature of the surface considered (K).   Typical values of emissivity for several materials are listed in Table 4.1.   From this table it is clear that emissivities for most materials involved in the present systems in this book are in the range between 0.9 and 1.0.   Emissivity does not depend on the color of the surface, and lustrous metal surfaces have very small values. Emissivity is not only a constant for emission of radiation but also a constant for absorption. The following two equations hold for most materials:

$$ALF + RMD + TAU = 1 \qquad (4.5)$$

$$\text{EPS} = \text{ALF} \qquad (4.6)$$

where **EPS** is emissivity, **RMD** is reflectivity, **TAU** is transmissivity, and **ALF** is absorptivity.

Typical values of each type of heat-flux component for the earth are shown in Fig. 4.2 as an annual budget for the earth, and are calculated based on the solar constant 1,360 W/m^2. Radiation exchange is often considered as a single process in which the terms of net radiation and effective radiation are used. Effective radiation on fine days (net in the night) is on the order of 100 W/m^2.

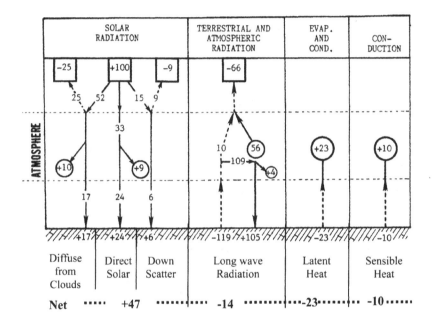

Figure 4.2. Energy exchange of the earth on annual balance (after Gates, 1962).

4.2. CONVECTIVE HEAT TRANSFER

The governing equation for convective heat transfer is shown as eq. 4.2, and eq. 4.3 is the equation for latent heat transfer, which will be described in the next section. The heat transfer coefficient, **H**, in eq. 4.2 is dependent on the movement of the adjacent air. Under outside conditions, wind speed is the primary factor, and the coefficient is expressed as a function of wind.

Wind speed consists of three directional flows -- **x**, **y**, and **z**, and the coefficient is related to the main directional flow, which is horizontal. The main horizontal wind speed changes with the logarithmic distance from the surface. It is rather difficult to determine the height at which wind speed should be taken.   However, if the air movement is completely turbulent, change is negligible beyond the boundary layer.   There is no rule for determining the height at present, but traditionally, anemometers are set 3 to 5 m above the ground surface.

Quite a number of experimental data have been obtained from wind tunnel experiments, and their results are summarized using non-dimensional numbers such as **Nu**, **Re**, **Gr** and **Pr**.   However, the situation in natural fields is different from that in a wind tunnel, and it would be useful to show the final relationship between the coefficient and the wind speed. Figure 4.3a summarizes several of the relationships reported, although they are used mostly for the outside surface of greenhouses.

In the present book, the following relationship is assumed throughout the text:

$$\mathbf{HO = CONS * V} \tag{4.7}$$

where **HO** is the heat transfer coefficient due to convection at the outside surface ($W/m^2/°C$), **CONS** is a constant, and **V** is wind speed (m/s).

In the range of free convection -- for example, under the film cover, between the cover and the soil surface, or in the greenhouse - - another expression is applied, where the heat transfer coefficient is a function of the temperature difference between the surface and the air.   The relationship between the heat transfer coefficient and temperature differences is summarized in Fig. 4.3b.   **MATLAB** scripts to draw Figures 4.3a and 4.3b are listed in Figure 4.4.

Again there are some discrepancies among the data, but in the present text, the following expression is assumed:

$$\mathbf{HI = CONS * DT} \tag{4.8}$$

where **HI** is the heat transfer coefficient due to free convection ($W/m^2/°C$) and **DT** is the temperature difference between the surface and the air ($°C$).

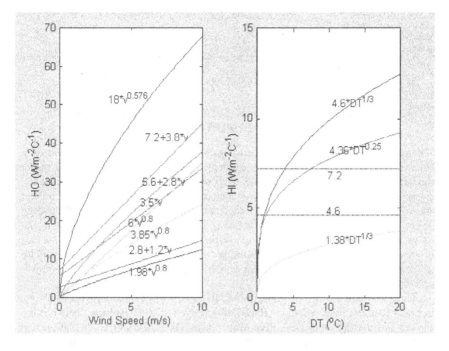

Figure 4.3a. Relationship between heat transfer coefficient and wind speed (after Takakura, 1989).

Figure 4.3b. Relationship between coefficient of free convection and temperature difference between the surface and the air (after Takakura, 1989).

```
%  Progam to draw Figures 4.3a and 4.3b.                    HOHI.m
clear all; clc;
subplot(1,2,1);
v=0:0.1:10;
v1=18*v.^0.576;     v2=7.2+3.8.*v;
v3=5.6+2.8.*v;      v4=3.5.*v;
v5=6*v.^0.8;        v6=3.85*v.^0.8;
v7=2.8+1.2.*v;      v8=1.98*v.^0.8;
plot(v,v1,v,v2,v,v3,v,v4,v,v5,v,v6,v,v7,v,v8);
xlabel('Wind Speed (m/s)');   ylabel('HO (Wm^{-2}C^{-1})');
axis([-inf inf 0 70]);
gtext('18*v^{0.576}');       gtext('7.2+3.8*v');
gtext('5.6+2.8*v');          gtext('3.5*v');
gtext('6*v^{0.8}');          gtext('3.85*v^{0.8}');
gtext('2.8+1.2*v');          gtext('1.98*v^{0.8}');
subplot(1,2,2);
dt=0:0.1:20;
y1=4.6*dt.^(1/3);            y2=4.36*dt.^0.25;
y3=7.2;                      y4=4.6;
y5=1.38*dt.^(1/3);
plot(dt,y1,dt,y2,dt,y3,dt,y4,dt,y5);
```

```
xlabel('DT (^oC)');                ylabel('HI (Wm^{-2}C^{-1})');
axis([-inf inf 0 15]);
gtext('4.6*DT^{1/3}');             gtext('4.36*DT^{0.25}');
gtext('7.2'); gtext('4.6');        gtext('1.38*DT^{1/3}');
```

Figure 4.4. MATLAB scripts to draw Figure 4.3 (**HOHI.m**).

4.3. A MODEL WITH SOLAR RADIATION AND AIR TEMPERATURE BOUNDARY CONDITION (**CUC02**)

The next model is a more sophisticated one which includes air temperature as a boundary condition and the radiation exchange and convective heat transfer at the soil surface. The outline of the model is shown in Fig. 4.5. In the present model, the soil layers are divided unevenly -- that is, thinnest at the surface and thicker toward the bottom, because the soil temperature does not change so much in deep layers. The top layer is 1 cm thick but is assumed to be a film surface in the balance equation. This assumption is justified because in practice the soil surface is not smooth and the surface temperature is not well defined and extremely difficult to measure correctly. This thickness can be reduced to 1 mm, for example, if necessary.

Three new components of heat transfer, all at the surface, are involved, as shown in Fig. 4.5: direct solar radiation (**RAD**), long wave radiation exchange between the surface and the sky, and convective heat transfer (**HO*(TF-TO)**). In the present model the atmospheric emissivity (**EPSA**) is assumed to be constant. The model is listed in Fig. 4.6, and its result is given in Fig. 4.7.

In the present model, the simulation model time clock **clk** is calculated by the **mod** function (see Fig. 4.6b). The variable **clk** therefore changes from 0 to 24 hours. Solar radiation (**RAD**) is calculated in '**solar.m**' in Fig. 4.6d.

The result of the simulation is shown in Fig. 4.7. The temperature boundary conditions are the same as in the preceding model except for air temperature. Although air temperature, one of the boundary conditions, ranges from 5 to 15°C in this case, the soil surface temperature is over 23°C because of solar radiation. The soil temperature range is expanded in the daytime and stays more or less the same in the nighttime as in the preceding model. It can be said that the change in soil temperature in the simulation is getting closer to the real pattern. Constants **RP** and **EPSA** can be changed to 100 and 0.71, respectively, in order to simulate severe radiation cooling in the night, and the results are shown in Fig. 4.7. Lower surface temperature than ambient air temperature in the nighttime is clearly shown, although the effect of the initial conditions remains for deeper soil layers.

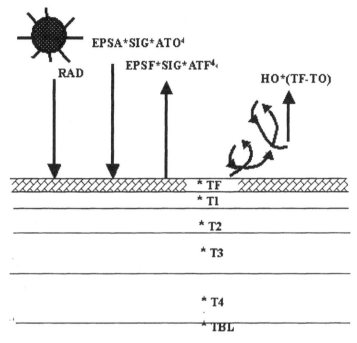

Figure 4.5. Diagram showing heat balance of soil layers.

```
%    Temperature regime in the soil layer                    CUC02.m
%    Boundary condition is air temp., Function required: soil02.m
clear all; clc
global RP EPSA
tic; RP=2000; EPSA=0.75; t0=0;  tfinal=48; y0=[10; 10; 10; 10; 10];
[t,y]=ode15s('soil02',[t0 tfinal],y0);    % Calling function 'soil2.m'
h1=findobj('tag','Temperature');close(h1);
figure('tag','Temperature','Resize','on','MenuBar','none',...
    'Name','CUC02.m (T in 5 soil layers given different RP & EPSA)',...
    'NumberTitle','off','Position',[160,80,520,420]);
subplot(2,1,1);
plot(t,y(:,1),'b+-',t,y(:,2),'b:',t,y(:,3),'b-.',t,y(:,4),'b--',...
    t,y(:,5),'b.-');
grid off;  axis([0, inf, 0, 25]);
tit=['RP=' num2str(RP) ' and EPSA=' num2str(EPSA)]; title(tit);
ylabel('Soil temperature, ^oC'); legend('TF','T1','T2','T3','T4',-1);
RP=100; EPSA=0.71; t0=0;  tfinal=48; y0=[5; 5; 6; 7; 8];
[t,y]=ode15s('soil02',[t0 tfinal],y0);
subplot(2,1,2);
h=plot(t,y(:,1),'k+-',t,y(:,2),'k:',t,y(:,3),'k-.',t,y(:,4),'k--',...
    t,y(:,5),'k.-');
set(h,'linewidth',2); grid off; axis([0, inf, 0, 10]);
tit=['RP=' num2str(RP) ' and EPSA=' num2str(EPSA)]; title(tit);
xlabel('time elapsed, hr'); ylabel('Soil temperature, ^oC');
legend('TF','T1','T2','T3','T4',-1);  toc
disp('Thank you for using');  disp(' ');
```

```
disp('CUC02: Program to calculate Temperatures in soil layers');
disp('      given various RP and EPSA.');disp(' ');
disp('You can enter ''close'' to close figure window.');
```

Figure 4.6a. Main program to simulate soil temperatures (**CUC02.m**).

Fig. 4.6a shows the main program of the **CUC02** model with the file name of '**cuc02.m**'. The commands listed are nearly the same as those in the main program of the **CUC01** model, and most statements are for setting constants and parameters. Fig. 4.6b shows the function program '**soil02.m**' which performs the main calculations for temperature regime in the soil layer and is similar to the subprogram '**soil01.m**' of **CUC01**. This program includes two more functions '**tabs.m**' and '**solar.m**', of which the scripts are listed in Fig. 4.6c and Fig. 4.6d, respectively. The function '**tabs.m**' calculates (absolute temperature/100)^4 and the function '**solar.m**' gives solar radiation **RAD** as a sine function and starts at 6 am and ends at 6 pm. The maximum value is given as **RP**, as the amplitude, but the negative values of the sine function are cancelled out by the "**if**" statement at the end.

```
%    Subprogram for cuc02 model                               soil02.m
%    Functions required: tabs.m and solar.m
function dy = soil02(t,y)
global RP EPSA
% RP: Solar radiation amp (kJ/m2/hr), EPSA: Emissivity of air layer
Tavg=10.0; TU=5.0; TBL=10.0; % Temp (C)
KS = 5.5; CS= 2.0E+3; HS = 25.2; SIG = 20.4;
% KS (kJ/m/C/hr) and KS/3.6 (W/m/C) also CS (kJ/m3/C)
% HS:Heat transfer coeff. at soil surface (kJ/m2/C/hr) = 7(W/m2/C)
% SIG:Stefan-Boltzmann constant (kJ/m2/K4/hr) = 5.67(W/m2/K4)
%      Please note that the above constant is 5.67e-8, the portion of 1e-8
%      has been calculated in function tabs() by (abs.T/100)^4.
Z0=0.01; Z1=0.05; Z2=0.1; Z3=0.2; Z4=0.5; % Depths of soil layer (m)
ALF = 0.7;     % ALF: Absorptivity of solar radiation at soil surface
EPSF = 0.95;   % EPSF:Emissivity of soil surface
clk = mod(t,24);   OMEGA=2.0*pi/24.0;  TO = Tavg + TU*sin(OMEGA*(clk-8));
TF=y(1);T1=y(2);T2=y(3);T3=y(4);T4=y(5);
TO4=tabs(TO);  TF4=tabs(TF);      % calling function tabs()
RAD=solar(RP,OMEGA,clk);          % calling function solar()
ITF = (ALF*RAD+EPSF*SIG*(EPSA*TO4-TF4)+ ...
         HS*(TO-TF)+KS*(T1-TF)*2.0/(Z0+Z1))/CS/Z0;
IT1 = (KS*(TF - T1)*2.0/(Z0+Z1)+KS*(T2 - T1)*2.0/(Z1+Z2))/CS/Z1;
IT2 = (KS*(T1 - T2)*2.0/(Z1+Z2)+KS*(T3 - T2)*2.0/(Z2+Z4))/CS/Z2;
IT3 = (KS*(T2- T3)*2.0/(Z2+Z3)+KS*(T4 - T3)*2.0/(Z3+Z4))/CS/Z3;
IT4 = (KS*(T3 - T4)*2.0/(Z3+Z4)+KS*(TBL - T4)*2.0/Z4)/CS/Z4;
dy=[ITF; IT1; IT2; IT3; IT4];
```

Figure 4.6b. Subprogram to simulate soil temperatures (**soil02.m**).

```
% Calculation for (absolute Temperature/100) to the power of 4    tabs.m
function TT4 = tabs(TT)
   TAA = (TT+273.16)/100.0;     TT4 = TAA*TAA*TAA*TAA;
```

Figure 4.6c. Subprogram to simulate soil temperatures (**tabs.m**).

```
% Subprogram for the calculation of solar radiation              solar.m
function rad = solar(RP,OMEGA,clk)
      rad= RP*sin(OMEGA*(clk-6.0));   % Sunrise at 6 am, sunset at 6 pm
      if rad<=0
          rad=0;
      end
```

*Figure 4.6d. Subprogram to simulate soil temperatures (**solar.m**).*

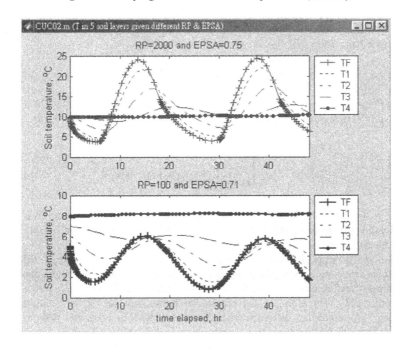

Figure 4.7. Top: Time courses of soil temperatures. Bottom: The effect of radiation cooling on soil temperatures.

4.4. MASS TRANSFER

Mass transfer occurs in the air as well as in the soil.   In the soil, not only water in two phases -- water and water vapor -- but also many kinds of salts move.   In the air, water vapor and carbon dioxide are two major transport components.   As water movement interacts with heat flow, a rather complicated model is required to represent the situation.   If the two flows are assumed independent, then modeling the situation is simplified.   The movement of salts is beyond the scope of this book. Water vapor and carbon dioxide movement in the air is rather important.   Water vapor flow from the soil surface to the adjacent air is always accompanied by energy flow.   Vaporization needs energy, as described in eq. 4.3.

Heat flow and water vapor flow can also be related through another number, the Lewis number, which is the ratio of thermal diffusivity and the molecular diffusion coefficient. The Lewis number is almost 0.96 for water vapor and 1.14 for carbon dioxide. Thermal diffusivity (**KAP**, m^2/hr) is expressed as

$$KAP = K / (RHO * CP) \qquad (4.9)$$

where **K** is thermal conductivity (kJ/m/hr/$^\circ$C), **RHO** is air density (kg/m^3) and **CP** is specific heat at constant pressure (kJ/kg/$^\circ$C). If concentrations are expressed on a volumetric basis, the vapor flux is expressed as

$$F = KMM * (PHII - PHIS) \qquad (4.10)$$

where **F** is vapor flux (kg/m^2/hr), **KMM** is the vapor flux coefficient (m/hr), and **PHII** and **PHIS** are the concentrations of vapor (kg/m^3). **KMM** is expressed as

$$KMM = 1 / \int_{zs}^{zi} (1/DD) \, dz = SH * DD / Z \qquad (4.11)$$

where **zs** and **zi** are distances from the soil surface where the concentrations are **PHIS** and **PHII**, respectively, **DD** is the molecular diffusion coefficient (m^2/hr) and **SH** is the Sherwood number. **zs** can be the soil surface.

The coefficient **KM** in eq. 4.3 presents some problems. In some books, **KM** is related to the molecular diffusion coefficient of the mass through another non-dimensional number, the Sherwood number, but it varies over a wide range due to wind speed. Therefore, in the present book, we will ignore the relationship between the molecular diffusion coefficient and **KM** and define **KM** as an independent constant.

The relationship between **KM** and **KMM** in eq. 4.3 is given as

$$KMM = KM / RHO \qquad (4.12)$$

The Lewis number, which is another non-dimensional number, is defined as

$$Le = 3.6 * H / (KM * CP) \qquad (4.13)$$

where **H** is the convective heat transfer coefficient and is the same as that used in eq. 4.2. If the Lewis number is equal to 1, the relation between heat flow and mass transfer is expressed as

$$KM = 3.6 * H / CP \qquad (4.14)$$

Since $CP = 1$ (kJ/kg/°C), then $KM = H * 3.6$ (kJ/m^2/hr/°C)/(kJ/kg/°C), that is (kg/m^2/hr). This relation is convenient because it is not necessary to give the mass transfer coefficient separately. The problem is to find the condition where $Le = 1$. It is known that Le changes with the ratio of relevant diffusivities and air velocity, and as air velocity over a wetted surface becomes low, Le decreases from 1 to 0.8 (Threlkeld, 1962). In the present book, Le is assumed to be 0.9 in all models. More details can be found in the book written by Threlkeld (1962).

4.5. A MODEL WITH LATENT HEAT TRANSFER (CUC03)

In heat transfer under normal conditions, normal air is wet and is defined as the mixed ideal gases of dry air and water vapor. It is not difficult to observe this by putting a glass of cold beer on a desk in the summer. The air around the glass is cooled and water vapor in the air condenses at the surface of the glass. This means that the air temperature at the surface of the glass is cooled below the dew-point temperature of the air at that temperature. This phenomenon can be shown on a psychrometric chart (Fig. 4.8).

The psychrometric chart is a powerful tool for understanding the physical processes of the air. Any condition of normal air can be expressed as a point in the area bounded by three lines: Temperature line (vertical lines), humidity ratio or water vapor pressure line (horizontal lines), and the curved saturation line (which is shifted according to the atmospheric pressure). The process of the air being cooled by a glass of beer is shown in the figure. The starting condition of the air is assumed to be x in the figure, and the cooling process is shown by the horizontal arrow pointed toward the saturation line. The cross point of this arrow and the saturation line is the dew-point, and its temperature is read by projecting it on the horizontal temperature scale. Suppose the surface temperature of the glass is lower than the dew-point indicated by the dot on the saturation line. Then the excess water vapor represented by the difference of the humidity ratio ΔW cannot stay in the air and is condensed. The humidity ratio shows the amount of water vapor per unit weight of dry air.

Fig. 4.9 shows the scripts for calculating temperatures in the soil layer when latent heat transfer is involved.

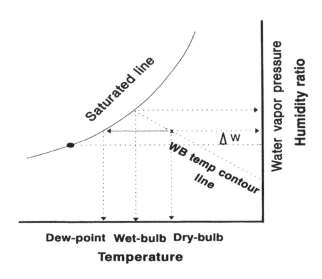

Dew-point Wet-bulb Dry-bulb
Temperature

Figure 4.8. Diagrammatic representation of psychrometric chart.

```
%    Temperature regime in the soil layer                      CUC03.m
%    Latent heat transfer is involved
%    Function involved: soil03.m
clear all;clc; tic
t0=0; tfinal=48; tfinal=tfinal*(1+eps);
y0=[10; 10; 10; 10; 10];   % ini. condition of TF, T1,T2,T3,T4
[t,y]=ode15s('soil03',[t0 tfinal],y0);    % Calling function soil3.m
h1=findobj('tag','Temperature'); close(h1);
figure('tag','Temperature','Resize','on','MenuBar','none', ...
    'Name','CUC03.m','NumberTitle','off','Position',[160,80,520,420]);
plot(t,y(:,1),'b^-',t,y(:,2),'gV-',t,y(:,3),'r+-', ...
    t,y(:,4),'c*-',t,y(:,5),'ko-');
grid on; axis([0, inf, 2, 20]);
title('EPSA=f(TD), also include the latent heat for evaporation');
xlabel('time elapsed, hr');
ylabel('Soil temperature, ^oC'); legend('TF','T1','T2','T3','T4',2);
toc; disp('Thank you for using ');    disp(' ');
disp('CUC03: Program to calculate Temperatures in soil layers');
disp('     given EPSA=f(TD), also include the latent heat for evaporation.');
disp(' '); disp('You can enter ''close'' to close figure window.');
```

*Figure 4.9a. Main program to simulate soil temperatures with latent heat exchange (**CUC03.m**).*

```
%  Subprogram for cuc03 model                                 soil03.m
%  Functions involved: FWS.m, TABS.m, SOLAR.m
%  rp=2000, EPSA=f(TD), also include the latent heat for evaporation
%  when calculating floor temperature
function dy = soil03(t,y)
Tavg=10.0; TU=5.0; TBL=10.0; % Temp (C)
TD=4.5;              %TD: outside dew point temperature, in degree C
KS = 5.5; CS= 2.0E+3; HS = 25.2;
% KS (kJ/m/C/hr) and KS/3.6 (W/m/C) also CS (kJ/m3/C)
```

```
% HS:Heat transfer coeff at soil surface (kJ/m2/C/hr), HS/3.6= 7 (W/m2/C)
rp=2000;          % rp: Solar radiation amp (kJ/m2/hr)
SIG = 20.4;       %SIG:Stefan-Boltzmann const. (kJ/m2/K4/hr) = 5.67(W/m2/K4)
Z0=0.01; Z1=0.05; Z2=0.1; Z3=0.2; Z4=0.5; % Depths of soil layer (m)
ALF = 0.7;        % ALF: Absorptivity of solar radiation at soil surface
EPSF = 0.95;      % EPSF:Emissivity of soil surface
HLG=2501.0;       LE=0.9;     KM=HS/LE;
WO=FWS(TD);       % calling function fws(), WO: Humidity ratio (outside air)
EPSA=0.711+(TD/100)*(0.56+0.73*(TD/100));% EPSA:Emissivity of air layer
clk = mod(t,24);   OMEGA=2.0*pi/24.0;     % Time (hr)
TO = Tavg + TU*sin(OMEGA*(clk-8));
TF=y(1);T1=y(2);T2=y(3);T3=y(4);T4=y(5);
TO4=tabs(TO);   TF4=tabs(TF);  %calling function tabs()
RAD=solar(rp,OMEGA,clk);   %calling func. solar() for simulated radiation
WF=fws(TF);                %calling function fws()
%WF: Humidity ratio at soil surface (assumed saturated)
ITF=(ALF*RAD+EPSF*SIG*(EPSA*TO4-TF4)+HS*(TO-TF)+HLG*KM*(WO-WF)+...
   KS*(T1-TF)*2.0/(Z0+Z1))/CS/Z0;
IT1 = (KS*(TF-T1)*2.0/(Z0+Z1)+KS*(T2-T1)*2.0/(Z1+Z2))/CS/Z1;
IT2 = (KS*(T1-T2)*2.0/(Z1+Z2)+KS*(T3-T2)*2.0/(Z2+Z3))/CS/Z2;
IT3 = (KS*(T2-T3)*2.0/(Z2+Z3)+KS*(T4-T3)*2.0/(Z3+Z4))/CS/Z3;
IT4 = (KS*(T3-T4)*2.0/(Z3+Z4)+KS*(TBL-T4)*2.0/Z4)/CS/Z4;
dy=[ITF; IT1; IT2; IT3; IT4];
```

Figure 4.9b. Subprogram to simulate soil temperatures with latent heat exchange (**soil03.m**).

```
% Calculate saturated humidity ratio                           FWS.m
function WWW=FWS(TTT)
Patm=101325; TQQ=TTT + 273.16;        T10=TQQ/100.0;
if TTT>0
       A=-5800.2206/TQQ+1.3914993-0.04860239*TQQ;
       B=0.41764768*T10*T10-0.014452093*T10*T10*T10;
       C=6.5459673*log(TQQ);
else
       A=-5674.5359/TQQ+6.3925247-0.9677843*T10;
       B=0.62215701E-2*T10*T10+0.20747825E-2*T10*T10*T10;
       C=-0.9484024E-4*T10*T10*T10*T10+4.1635019*log(TQQ);
end
BETA=A + B + C;       PWS=exp(BETA);       WWW=0.622*PWS/(Patm-PWS);
```

Figure 4.9c. Subprogram to simulate soil temperatures with latent heat exchange (**fws.m**).

In this model (**CUC03**), the boundary condition at the surface was modified to be close to that in reality; that is, latent heat transfer has been included in this model. The rest of the model from section 4.3 is unchanged (see Fig. 4.1). Evaporation from the soil surface is now in the model and the latent heat transfer (**HLG*KM*(WF-WO)**) has been added. The humidity ratio of the soil surface is assumed saturated at the soil surface temperature and is calculated by using the saturation curve in Fig. 4.8 (see also ASHRAE, 1985).

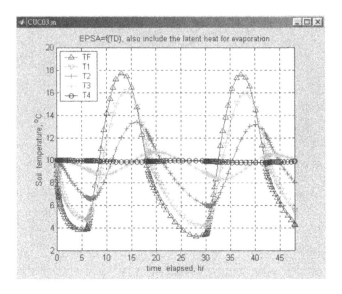

Figure 4.10. Simulated results of CUC03 model.

The structure of the model is similar to the one in section 4.3. In the present program, there are six unknown variables, **TF, T1 - T4**, and **WF**; and 6 equations.

$$\mathbf{TF} = f_1(\mathbf{TF, T1, Wf}), \qquad \mathbf{T1} = f_2(\mathbf{TF, T1, T2}), \qquad \mathbf{T2} = f_3(\mathbf{T1, T2, T3})$$
$$\mathbf{T3} = f_4(\mathbf{T2, T3, T4}), \qquad \mathbf{T4} = f_5(\mathbf{T3, T4}), \qquad \mathbf{WF} = f_6(\mathbf{TF})$$

In the present model shown in Fig. 4.9, the new function **FWS** in subprogram 'fws.m' (see Fig. 4.9c) for calculating the saturated humidity ratio is included. This semi-empirical expression calculates saturated humidity ratio (dummy argument is **WWW**) if a wet-bulb temperature (dummy argument is **TTT**) is given (after ASHRAE, 1988). Other approximations of the curve are also available with less accuracy (see **CUC150**).

One of the new boundary conditions is dew-point temperature **(TD)** of the air. Using **TD**, humidity ratio of the air **(WO)** as well as atmospheric emissivity **(ESPA)** are calculated by eq. 4.18. Humidity ratio at the soil surface **(WF)** is also calculated using the function **FWS**.

The results of the model are given in Fig. 4.10. In the present case, the soil is assumed to be completely wet, that is, saturated at the soil surface. Therefore, potential evaporation takes place and cools down the soil surface temperature. The maximum temperature of the surface is just above 18°C, which is 5°C lower than in the case presented in section 4.3. Deeper soil layers have little effect on the surface.

4.6. RADIATION BALANCE

Radiation is emitted from a body whose temperature is above 0 K; in the present situation, that is, everything emits radiation. Radiation is always considered a balance. For the radiation balance of the bare ground, net radiation is defined as the difference between incoming radiation from the sky and outgoing radiation from the ground. In the daytime, the balance includes solar radiation, of course. The net radiation only for long wave radiation is called effective radiation. In the nighttime, net radiation is equal to effective radiation.

Radiation is like light; it can be considered to travel in a straight line. If two bodies, such as the sky and the ground, are placed face to face without any intermediary, the radiation exchange between these two bodies is one to one. The heat balance systems we have so far considered are of this type -- that is, the bodies involved are the flat soil surface and the sky. These two surfaces can be considered infinite and parallel. Therefore, the radiation emitted from one surface always enters the other.

But when another body such as a building is involved, as shown in Fig. 4.11, part of the radiation from the ground cannot go directly to the sky, and vice versa. This situation can be visualized by looking at the figure from the direction shown by the word 'eye'. Suppose the building is infinite in the direction perpendicular to the page: then half of the sky and half of your perspective are occupied by the face of the building. The ground does receive half of the radiation from the sky, but it also receives half of that from the building surface. In this case, it is said that the view factor of the ground to the sky is 0.5 and the view factor of the ground to the building is 0.5. On the other hand, if you move your eye from the ground to the building surface, it is clear that the view factor of the building to the sky is 0.5. Of the three items being discussed, normally the sky has the lowest temperature and the smallest amount of long wave radiation. On calm clear nights, the soil surface is cooled by radiation because of this fact. However, if half of the sky is blocked by a large building such as shown in Fig. 4.11, the soil surface will be warmer because more radiation will travel from the building surface to the soil surface. Therefore, it is very important to take into account view factor relations when we analyze radiation exchange among surfaces of different surface temperatures.

View factor takes into consideration the area of another body with which the radiation exchange takes place in relation to the total hemisphere of the exchange surface. This relationship is three-dimensional, and the geometric relationship between the two bodies is not simple. A simple way to calculate the view factor is to project all the areas onto the horizontal surface with which the basic surface is involved; then the hemisphere becomes the unit circle and the projected area of the body is the view factor, as shown in Fig. 4.12. The important relationship is that

$$\sum_{j=1}^{n} \text{view factor}(\mathbf{i}, \mathbf{j}) \quad = 1 \qquad\qquad (4.15)$$

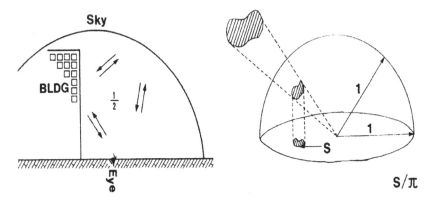

Figure 4.11. Radiation exchange among   Figure 4.12. Geometrical representation of
      the sky, the ground and a building.   view factor.

which means that the summation of the view factors of the arbitrary surface **i** to any surface **j** including itself is equal to 1. The self view factor is defined when the surface is concave.

If we look at protected cultivation systems closely, most parts of the systems are not flat surfaces. Rows, tunnels and houses all have curved surfaces of different temperatures. Therefore, it is important to consider view factors in detailed analyses of these systems. However, in the present book, in order to simplify the problem, view factors are not taken into account in any model; it is assumed that all surfaces are infinitely flat. View factor analyses have been conducted in some models (*e.g.*, Takakura *et al.*, 1971), which can be consulted when necessary.

Table 4.1. Emissivities of various surfaces (after Seller, 1965).

A. *Water and Soil Surfaces*		C. *Vegetation*	
Water	0.92-0.96	Alfalfa, dark green	0.95
Snow, fresh fallen	0.82-0.995	Oak leaves	0.91-0.95
Ice	0.96	Leaves and plants	
Sand, dry light	0.89-0.90	0.8 µm	0.50-0.53
Sand, wet	0.95	1.0 µm	0.50-0.60
Gravel, coarse	0.91-0.92	2.4 µm	0.70-0.97
Limestone, light gray	0.91-0.92	10.0 µm	0.97-0.98
Concrete, dry	0.71-0.88	D. *Miscellaneous*	
Ground, moist, bare	0.95-0.98	Paper, white	0.89-0.95
B. *Natural Surfaces*		Glass, pane	0.87-0.94
Desert	0.90-0.91	Bricks, red	0.92
Grass, high dry	0.90	Plaster, white	0.91
Fields and shrubs	0.90	Wood, planed oak	0.90
Oak woodland	0.90	Paint, white	0.91-0.95
Pine forest	0.90	Paint, black	0.88-0.95
		Paint, aluminum	0.43-0.55
		Aluminum foil	0.10-0.5
		Iron, galvanized	0.13-0.28
		Silver, highly polished	0.20
		Skin, human	0.95

4.7. LONG WAVE RADIATION

Radiation is emitted from any body whose temperature is above 0 K, including the sky and the earth's surface. Radiation emitted from a body whose temperature is in the ordinary range $(0 - 30\,°C)$ has a peak of density in the longer wavelengths shown in Fig. 4.13 called long wave radiation.

Emissivity and temperature of the atmosphere are not simple to determine; neither is long wave radiation from the sky. Within a rather thin layer such as 10 m or so, it can be assumed that the air is transparent to long wave radiation. This assumption, however, cannot be applied to the whole layer of the atmosphere around the earth. Atmospheric radiation is a function of water vapor, carbon dioxide and ozone content, but in most cases, the effects of carbon dioxide and ozone are neglected, mainly because of their thin, very low emissivities.

A large number of radiation charts have been developed for computing atmospheric radiation (*e.g.*, Seller, 1965). They need data that are not available, and the techniques for computing the radiation are not simple. Empirical equations, which are less accurate, are more practical. Brunt's equation (in van Wijk, 1966) is

$$RD = (A + B * sqrt(WO)) * SIG * ATA^4 \qquad (4.16)$$

where **RD** is downward long wave radiation (atmospheric radiation) (W/m^2), **A** and **B** are empirical constants related to each location with values of 0.53 and 0.06, respectively, for various locations in U.S.A., **WO** is humidity ratio (kg/kg DA), and **ATA** is absolute air temperature (K) at the screen height. Since **A** and **B** are dependent on temperature and water vapor of the atmosphere, these constants must be determined at the place where this equation is applied.

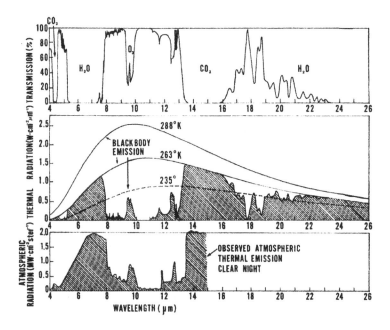

Figure 4.13. Long wave radiation from the sky (after Gates, 1962).

A similar empirical equation was reported based on observation at six locations in the U.S.A. by Martin and Berdahi (1984):

$$RD = EPSA * SIG * ATA^4 \qquad (4.17)$$

$$EPSA = 0.711 + 0.56 * TD / 100 + 0.73 * (TD / 100)^2 \qquad (4.18)$$

where **EPSA** is atmospheric emissivity and **TD** is dew-point temperature (°C) at the screen height. Emissivities of various materials are summarized in Table 4.1.

MATLAB FUNCTIONS USED

clc	Clear the Command Window and home the cursor.
clear	Clear variables and functions from memory.
clear all	Remove all variables, globals, functions and MEX links.
exp	Exponential. **exp**(X) is the exponential of the elements of X.
for.... end	Repeat statements a specific number of times. The general form is: **for** variable = expr, statement, ..., statement **end**
gtext	Place text with mouse. **gtext**('string') displays the graph window, puts up a cross-hair, and waits for a mouse button or keyboard key to be pressed.

if..else..end	IF statement condition. The general form is:
	if expression
	    statements
	    elseif expression
	    statements
	    else
	    statements
	    end

log	The natural logarithm of the elements of X.

arithmetic	+	Plus	-	Minus
operators	*	Matrix multiply	.*	Array multiply
	^	Matrix power	.^	Array power
	\	Backslash or left matrix divide	/	Slash or right matrix divide
	.\	Left array divide	./	Right array divide

relational	==	Equal
operators	<	Less than
	<=	Less than or equal
	~=	Not equal
	>	Greater than
	>=	Greater than or equal

logical	&	Logical AND
operators	~	Logical NOT
	|	Logical OR
	xor	Logical EXCLUSIVE OR

PROBLEMS

1. If you look carefully at the periodic changes of the deeper soil layers in Fig. 4.7, a trend of temperature rise or fall is clear. Explain the reason for this, then change the initial conditions for all soil layers to more reasonable values. Note: Two days' run is not enough to check this. Run for a week at least.

2. Modify the program **CUC02**, applying the convective heat transfer coefficient at the soil surface (**HS**) as a function of wind speed. Use the following relationships: **HS** = 2.8 + 1.2 * **V** and **V** = 2.0 + 2.0 * **sin** (**OMEGA** * **clk**).

3. Variable **PWS** (in Pa) in the function **FWS.m** calculates saturated vapor pressure as a function of given temperature (see Fig. 4.9c). Function **VAPRES** (in mb) in the **CUC151** program (see Chapter 10) uses a different approximation. Compare these two methods of calculating saturated vapor pressure in the temperature range of $0 - 40°C$.

4. Explain the reason why the front glass of a car facing the sky on a winter night is frosted, while that of a car facing a tall building is not frosted.

5. Calculate **RD** in eqs. 4.16 and 4.17 for the dew-point temperature range $0 - 20°C$ and air temperature assumed 30 degree C. Program '**FWS.m**' is required to derive humidity ratio from dew-point temperature.

6. Modify the model **CUC01**, assuming the thermal conductivity of soil **KS** is not a constant and is a function of soil temperature (**TEMP**), expressed as: **KS** = 5.5 + 0.1 * **TEMP**.

7. Modify the model **CUC03**, assuming the temperature of the sky follows the Swinkbank model. The expression is: $Tsky = 0.0552 * (TO)^{1.5}$. Both Tsky and TO are expressed in Kelvin. Use Tsky for the calculation of the long wave radiation from the sky.

CHAPTER 5

SOLAR RADIATION ENVIRONMENT

5.1. INTRODUCTION

Solar radiation is one of the most important environmental factors for plant growth. In general it is known that radiation is generated from a material close to a blackbody. Solar radiation received at the earth's surface varies with the season because of the planetary relation between the sun and the earth. Therefore, for cultivation in controlled environments, it is very important to calculate how much of the solar radiation we can utilize at a given place on the earth at a given time of the year.

Coverings can improve the temperature environment by increasing inside temperature, but they cannot enhance the solar radiation level. Shading to reduce solar radiation level is sometimes important, but in most cases, the most important problem is minimizing the reduction of solar radiation due to coverings.

Three properties of coverings are related to the solar radiation environment: transmissivity, reflectivity and absorptivity. It is hard to see with the human eye, but solar radiation is decreased by reflectivity and absorptivity as it travels through transparent films. In a normal situation, the amount of solar radiation transmitted into the covered area is less than that outside.

These three properties are dependent on wavelength. PVC film in the early years created serious problems for eggplant production under PVC cover, because the stabilizer in the film absorbed solar radiation in the ultraviolet region (290 - 360 nm) that triggers the violet coloring of eggplants. Another problem is that the eyes of honeybees are sensitive in the ultraviolet region (< 400 nm), and the bees cannot fly well if the covering does not transmit solar radiation in this region. Therefore, it is essential to be careful about the spectral distribution of transmitted solar radiation if you are growing plants with color generated by anthocyanin (a pigment for red to violet color) or plants with flowers pollinated by bees.

5.2. UNITS OF RADIATION AND LIGHT

The radiation in the range from 400 to 700 nm is called photosynthetically active radiation (PAR) for plants. The photoreactive part of photosynthesis is considered directly proportional to the amount of photons in the light which have been absorbed by the plant in accordance with the chlorophyll absorptance curve when CO_2 concentration is not limiting. This curve shows the relationship of photons absorbed as a function of wavelength and has two peaks, in the blue and red regions. A mole of photons is 6.02×10^{23} photons, with each photon having energy proportional to its frequency. The photon energy of green light whose wavelength is 500 nm is

65

approximately 2.35 x 10^5 J/mol photon (*e.g.*, Takakura, 1991). For the range of PAR (400-700 nm), a conversion factor of 4.6 μmol photon /J can be used (Ting and Giacomelli, 1987).

Light, the visible part of radiation perceived by the human eye, is commonly measured by a lux meter, an instrument which has a filter of the range from 400 to 700 nm and a sharp peak in the green region. The amount of illumination is expressed in a unit called lux (lx). Plants have different absorption characteristics than the human eye, and lux cannot be correctly used as a measure of the light for a plant. Light measured by lux meters will exaggerate the green light available. Radiometers which have a linear filter with transmissivity slightly greater in red than blue wavelengths, are also used to measure PAR (μmol photon/m^2/s or W/m^2). If radiation is measured beyond the visible region, an energy unit such as W/m^2 is used. If an instrument has a particular filter, such as a lux meter, its readings are restricted to be used for a special purpose. If universal comparison among different radiation sources, such as radiation from the sun and from various artificial light sources, is necessary, and photosynthesis is not a consideration, then an energy unit such as W/m^2 would be better, even for the visible region. The results should clearly indicate the wavelength measured.

The annual average of the irradiance measured outside of the earth's atmospheric layer on a surface held perpendicular to the solar beam is called the solar constant. Its value is 1367 ± 7 W/m^2 (see eq. 5.10). The solar constant varies slightly according to the sun's activity as well as the distance between the sun and the earth. The solar energy flux density measured at sea level cannot exceed this value due to attenuation by the atmosphere. A maximum of approximately 75% of the solar constant or 1000 W/m^2 can be measured at the Earth's surface (Monteith and Unsworth, 1990).

Near noon on a clear day the average photon flux density on the ground in PAR can be calculated; assuming a measured total solar flux density of 500 W/m^2, then multiplying by 4.6 μmol photon/J will result in 2300 μmol photon/m^2/s. Due to the reason that PAR range (400-700 nm) only occupies 38.15 % of the total solar energy (Ting and Giacomelli, 1987), 877 μmol photon/m^2/s of PAR can be derived. PAR flux density from artificial light sources such as high pressure sodium lamps and fluorescent lamps, commonly used in practice, varies from 200 to 1000 μmol photon/m^2/s.

5.3. SOLAR RADIATION PROPERTIES OF COVERING MATERIALS

The spectral distributions of transmissivity for three types of films are shown in Fig. 5.1. Spectral distribution of transmissivity is important, but reflectivity and absorptivity should also be noted. All three properties are expressed in either decimal fractions or percentages; their total is 1, or 100%. If expressed in decimal fractions,

$$\textbf{TRAN} + \textbf{REFL} + \textbf{ABSO} = 1 \qquad (5.1)$$

where **TRAN** is transmissivity, **REFL** is reflectivity and **ABSO** is absorptivity, all in decimal fractions. This equation holds for both monochromatic and broad spectrum of wavelength.

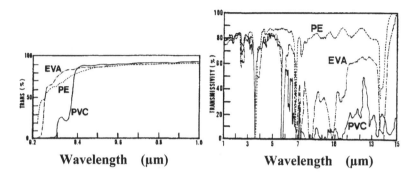

Figure 5.1. Spectral distribution of transmissivity of films (after Takahashi, 1975).

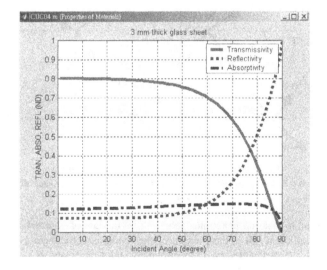

Figure 5.2. Transmissivity, reflectivity and absorptivity of a 3 mm glass sheet with an index of refraction of 1.526.

As we discussed, transmissivity is the most important of the three properties. It depends on the incident angle of solar radiation to the film, as does reflectivity. Reflectivity increases while transmissivity decreases with an increase in the incident angle, as shown in Fig. 5.2. Absorptivity is fairly constant through all incident angles from 0 to 90 degrees.

5.4. CALCULATION OF TRANSMISSIVITY (**CUC04**)

Although the following relationships all hold monochromatically -- that is, based on unit wavelength -- all representations here will be of total energy for the sake of simplicity.

As light hits a transparent material such as PVC film, direct solar radiation is partly transmitted and partly reflected and absorbed. The detailed mechanism is shown in Fig. 5.3. An incoming light ray **J** (incident angle **THET**) is partly reflected at the surface and is refracted (refracted angle **THETP**). We have the relation

$$\mathbf{FN = sin(THET) / sin(THETP)} \qquad (5.2)$$

where FN is the index of refraction of a film. The index of refraction of the air is unity.

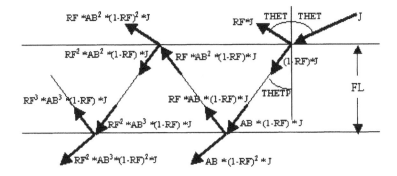

Figure 5.3. Multiple reflections of direct solar radiation by a sheet of film (after Threlkeld, 1962).

The ray in the material penetrates according to the law of Lambert-Beer: that is, absorption coefficient **AB** is the fraction of the radiation component available after absorption, also termed single pass transmittance, and is expressed as

$$\mathbf{AB = exp(-FK * FL / cos(THETP))} \qquad (5.3)$$

where **FK** is the extinction coefficient (1/mm) and **FL** is the thickness of the film (mm). This equation is the expression of Lambert-Beer's law and the term **FL/cos(THETP)** is the actual path length for the radiation beam. As shown in Fig. 5.3, because of successive internal reflections, the equations for reflected, absorbed and transmitted light are given by the sums of infinite series. Let **RF** be the fraction of each component reflected, termed specular reflectance or single pass reflectance. The total transmissivity **TRAN** is given by

$$\text{TRAN} = (1 - \text{RF})^2 * \text{AB} / (1 - \text{RF}^2 * \text{AB}^2) \tag{5.4}$$

In a similar way, we obtain the total reflectivity **REFL** as

$$\text{REFL} = \text{RF} + \text{RF}*(1-\text{RF})^2*\text{AB}^2/(1-\text{RF}^2*\text{AB}^2) = \text{RF}*(1+\text{TRAN}*\text{AB}) \tag{5.5}$$

and absorptivity **ABSO** as

$$\text{ABSO} = 1 - \text{RF} - (1 - \text{RF})^2 * \text{AB} / (1 - \text{RF} * \text{AB}) = 1 - \text{TRAN} - \text{REFL} \tag{5.6}$$

The reflectivity component **RF** may be derived from the Fresnel relations. Light is a kind of electromagnetic wave, and natural or unpolarized light may be assumed to consist of two vibrating components, one vibrating in a plane normal to the sheet and the other vibrating in a plane parallel to the sheet. A light ray entering a flat material reflects at the first surface and again at the second surface. The reflectivity component, which is the ratio of the reflected part to the entering light is

$$\text{RP} = \tan^2 (\text{THET} - \text{THETP}) / \tan^2 (\text{THET} + \text{THETP}) \tag{5.7}$$

in the plane parallel to the sheet and

$$\text{RN} = \sin^2 (\text{THET} - \text{THETP}) / \sin^2 (\text{THET} + \text{THETP}) \tag{5.8}$$

in the plane normal to the sheet. If the reflected components are of equal intensity in the parallel and normal planes,

$$\text{RF} = (\text{RN} + \text{RP}) / 2 \tag{5.9a}$$

If **THET** equals 0, **RF** is calculated using the follow equation:

$$\text{RF} = (1 - 1 / \text{FN})^2 / (1 + 1 / \text{FN})^2 = (\text{FN} - 1)^2 / (\text{FN} + 1)^2 \tag{5.9b}$$

The program to obtain the result shown in Fig. 5.2 is given in Fig. 5.4. In eq. 5.9a, incident angle **THET** is given, but **THETP** is an unknown variable. Therefore, eq. 5.9 is broken down into its components: that is, the trigonometric functions for **THET** and **THETP** are separated, and then the trigonometric function for **THETP** is found from eq. 5.2 because the value of **FN** is also given. **Cos(THETP)** is also calculated from eq. 5.2. **AB** is calculated by eq. 5.3, as **FK** is given. Finally **TRAN** is calculated by eq. 5.4, and **REFL** and **ABSO** from eqs. 5.5 and 5.6, respectively. In the program given in Fig. 5.4b, several intermediate variables for trigonometric functions used for convenience such as **STHET**, **CTHET**, **STHETP**, **CTHETP**, **STHENG**, **CTHENG**, **STHEPS**, **CTHEPS**, **TANPS**, **TNANNG**, **STHEN2**, **TANPS2** and **TANNG2** might make the program difficult to follow, but the explanation given here will help us to understand the program.

As clearly shown in Fig. 5.2, transmissivity drops rather rapidly after the incident angle exceeds 60 degrees. On the other hand, absorptivity is fairly constant throughout the range, and it is around 12% in this case. Therefore, the decrease in transmissivity occurs along with the increase in reflectivity. The values of **FL**, **FN** and **FK** are material dependent. Table 5.1 shows these properties of some popular glazing materials. In Table 5.1, information of **FL** and **FN** of selected glazing were compiled from the commercial catalogue. Extinction coefficients **FK** were calculated based on the equation, derived from eqs. 5.3 and 5.4, listed below (Fang, 1992).

$$\mathbf{FK} = -\log\left((\mathbf{X}^{0.5} - \mathbf{B}) / (2 * \tau * \mathbf{RF}^2)\right) / \mathbf{FL} \tag{5.9c}$$

where $\mathbf{X}$ equals $(\mathbf{B}^2 + 4 * \tau^2 * \mathbf{RF}^2)$, $\mathbf{B}$ equals $(\mathbf{RF}-1)^2$, $\mathbf{RF}$ equals $[(\mathbf{FN}-1)/(\mathbf{FN}+1)]^2$, τ is the direct transmissivity when incident angle (**THET**) equals zero and **FL** is the thickness of glazing (in mm).

```
% Program to calculate optical properties of glazing        CUC04.m
% Function required: RFcal.m
clear all;clc
tit='3 mm thick glass sheet';
FL=3;        % Thickness (mm)
FN=1.526;    % FN: Index of refraction
FK=0.0441;   % FK: extinction coefficient (1/mm)
% Above 3 values are subject to change for different glazing.
TRAN=zeros(1,90);REFL=zeros(1,90);ABSO=zeros(1,90);
for angle=1:90                     % ANGLE: Incident angle in degree
    THET=angle*pi/180;             % THET: Incident angle in radian
    [RF, CTHETP]=RFcal(THET,FN);   % Calling function RFcal()
    AB=exp(-FK*FL/CTHETP);
    TRAN(angle)=(1-RF)*(1-RF)*AB/(1-RF*RF*AB*AB);
    REFL(angle)=RF+RF*(1-RF)*(1-RF)*AB*AB/(1-RF*RF*AB*AB);
    ABSO(angle)=1-RF-(1-RF)*(1-RF)*AB/(1-RF*RF*AB);
end
% TRAN: Transmissivity, REFL: Reflectivity and ABSO: Absorptivity
h1=findobj('tag','Coefficients'); close(h1);
figure('tag','Coefficients','Resize','on','MenuBar','none',...
   'Name','CUC04.m (Properties of Materials)','NumberTitle','off',...
   'Position',[200,40,520,420]);
x=1:1:90;
plot(x,TRAN(:),'r-',x,REFL(:),'b:',x,ABSO(:),'k-.','linewidth',4);
axis([0,inf,0,1]);grid on; title(tit);
xlabel('Incident Angle (degree)'); ylabel('TRAN, ABSO, REFL (ND)');
legend('Transmissivity','Reflectivity','Absorptivity');
disp('You can enter ''close'' to close figure window.');
%END
```

*Figure 5.4a. Main program to calculate light properties of a film (**CUC04.m**).*

Table 5.1. Properties of some popular glazing (after Fang, 1994).

Glazing	FL Thickness	FN, Index of refraction	FK, Extinction coefficient (1/mm)
Acrylic,(Acrysteel*,1/8 inch)	3.175 mm	1.56	0.0065
EVA	0.15 mm	1.515	0.0699
FRP	25 mil**	1.54	0.2482
Glass (ordinary float)	3.175 mm	1.526	0.0473
Glass (double strength)	3.175 mm	1.526	0.0094
Glass (sheet lime)	3.175 mm	1.51	0.0178
PC (Lexan*)	3.175 mm	1.59	0.0662
PC (Lexan dripguard*)	6 mm	1.586	0.0042
PE (UV resistant)	4 mils	1.515	0.0752
PE (IR barrier, Monsanto 602*)	4 mils	1.515	0.432
PE	4 mils	1.515	0.165
Polyester (Mylar*)	5 mils	1.54	0.205
PVC (Bioriented*)	0.9 mm	1.46	0.17
PVC (clear)	0.15 mm	1.46	0.09
PVC (haze)	0.15 mm	1.46	0.3106
PVF (Tedlar*)	2 mils	1.46	0.4806

* Trade names   ** 1 mil is 1/1000 inch

```
% Function to calculate fraction of each component reflected        RFcal.m
function [RF, CTHETP]=RFcal(THET,FN)
%
STHET=sin(THET);   CTHET=cos(THET);
if (THET-0.0)<=0.000001
        RF=(1-1/FN)*(1-1/FN)/((1+1/FN)*(1+1/FN));
        CTHETP = 1;
else
        STHETP=STHET/FN;    CTHETP=sqrt(1.0-STHETP*STHETP);
        STHENG=STHET*CTHETP-CTHET*STHETP;
        STHEPS=STHET*CTHETP+CTHET*STHETP;
        CTHEPS=CTHET*CTHETP-STHET*STHETP;
        CTHENG=CTHET*CTHETP+STHET*STHETP;
        TANPS=STHEPS/CTHEPS;         TANNG=STHENG/CTHENG;
        STHEN2=STHENG*STHENG;        STHEP2=STHEPS*STHEPS;
        TANPS2=TANPS*TANPS;          TANNG2=TANNG*TANNG;
        RF=0.5*(STHEN2/STHEP2+TANNG2/TANPS2);
end
```

Figure 5.4b. Function used in **CUC04** *model* **(RFcal.m)**.

5.5. SOLAR RADIATION

The sun is the largest energy source in our system; we get an equivalent temperature of 5,700 K if we calculate its temperature from its radiation, as shown in Fig. 5.5. The solar constant **J0W** (defined in section 5.2) varies, as it is a kind of indicator of the energy that reaches the earth, based on observations over many years. It is reported that the value of the solar constant throughout the year can be calculated using the following equation:

$$J0W = Jsc * (1 + 0.033 * cos(2 * pi * n / 365)) \qquad (5.10)$$

where **n** is the Julian day, **Jsc** was equal to 1353 W/m^2, which is the average of **J0W** throughout the year. The value was updated to 1367 ± 7 W/m^2 in 1981.

When solar radiation passes through the atmosphere around the earth, some parts of the radiation are absorbed, and others reflected (see Fig. 4.2). The solar radiation received on the earth's surface, therefore, consists of directional and non-directional radiation -- that is, direct and diffuse radiation. Direct radiation is that which comes from the sun directly, and diffuse is that which is reflected in the atmosphere and comes from all directions of the sky. Hemispherical radiation on a horizontal surface is given as the sum of two components, **RAD** and **RADS**, where **RAD** is direct radiation (W/m^2) and **RADS** is diffuse radiation (W/m^2) (see Fig. 5.6).

Sometimes it is convenient to use semi-empirical mathematical expressions to calculate these two radiation components. The equation for **RAD** is

$$RAD = J0W * sin(SALT) * PP^{\wedge}(1 / sin(SALT)) \qquad (5.11)$$

where, **J0W** is solar radiation irradiated to a surface normal to sun ray outside the atmosphere (W/m^2), and is derived from eq. 5.10. **SALT** is the sun's altitude and **PP** (ND) is atmospheric transmittance. Extinction of **J0W** due to the air layer is counted. The direct radiation is directional; therefore, the so-called cosine law rules:

$$J0W * sin(SALT) = J0W * cos(THET) \qquad (5.12)$$

where **THET** is the incident angle of the radiation to the horizontal surface on earth. The diffusion of solar radiation due to the air layer is calculated as follows:

$$RADS=J0W*sin(SALT)*(1-PP^{\wedge}(1/sin(SALT)))/(1-1.4*log(PP))/2 \quad (5.13)$$

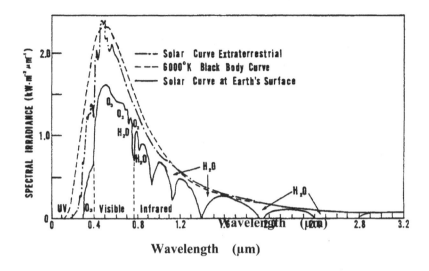

Figure 5.5. Solar radiation (after Gates, 1962).

An example of the calculated direct and diffuse solar radiation using eqs. 5.11 and 5.13 is shown in Fig. 5.6. The parameter **PP** is atmospheric transmittance (also termed clearness index), which gives the degree of sky clearness: the larger **PP**, the clearer the sky. Therefore, when the value of **PP** is large, the portion of direct solar radiation is large and the diffuse portion is small and vice versa as shown in Fig. 5.6b. The program, **CUC04a.m**, to generate Fig. 5.6b is listed in Fig. 5.6a.

```
% Program to calculate solar radiation on the earth          CUC04a.m
%
clear all;clc
h1=findobj('tag','cuc04a');  close(h1);
prgtitle='CUC04a.m (Direct and diffuse '
prgtitle= prgtitle+'solar radiation under various atmospheric transmittance)';
figure('tag','cuc04a','Resize','on','MenuBar','none','Name', ...
    prgtitle, 'NumberTitle','off','Position',[200,80,520,420]);
%
J0W=1367; % Solar Constant in W/m2
conv=pi/180; RAD=zeros(3,90); RADS=zeros(3,90);
for j=1:3
    pp=0.5+0.2*(j-1);  % pp=0.5,0.7,0.9
    for xx=1:90
        salt=xx*conv;  ALTS = sin(salt); ALTC = 1/ALTS;
        ppp=pp^ALTC;  RAD(j,xx)=J0W*ALTS*ppp;              % eq.5.11, in W/m2
        RADS(j,xx)=J0W*ALTS*(1-ppp)/(1-1.4*log(pp))/2;     % eq.5.13, in W/m2
    end
end
x=1:1:90;
plot(x,RAD(1,:),'r-',x,RAD(2,:),'b:',x,RAD(3,:),'k-.',x,RADS(1,:), ...
```

```
'r-',x,RADS(2,:),'b:',x,RADS(3,:),'k-.','linewidth',2);
axis([0,90,0,1400]);
xlabel('Solar Altitude, degree'); ylabel('Solar Radiation, W/m^2');
legend('Direct,  PP=0.5','Direct,  PP=0.7','Direct,  PP=0.9',...
    'Diffuse, PP=0.5','Diffuse, PP=0.7','Diffuse, PP=0.9',2);
title('Direct and Diffuse Solar Radiation');
text(61,900,'Direct'); text(61,220,'Diffuse');
grid on;
disp('You can enter ''close'' to close figure window.');
```

*Figure 5.6a. Program to calculate solar radiation on the earth (**CUC04a.m**).*

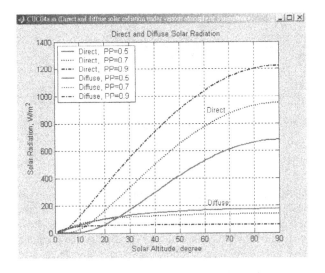

Figure 5.6b. Solar radiation on the earth calculated using eqs. 5.11 and 5.13.

5.6. THE SUN'S ALTITUDE AND AZIMUTH

5.6.1. On a horizontal surface

It is clear that the sun's position is relative to the place and time at which it is observed. The sun's position is expressed in terms of the sun's altitude (**SALT**) and azimuth (**SAZM**) as shown in Fig. 5.7. They are expressed as angles (degrees). For altitude, the angle is taken from the horizontal plane upward toward the sun with the upward direction as positive -- that is, -90 < **SALT** < 90, and **SALT** = 0 at the horizontal surface. The angle for the azimuth is taken from the south when angles toward the west are positive -- that is, -180 < **SAZM** < 180 and **SAZM** = 0 at solar noon (when the sun is located to the exact south).

The observation location is expressed in terms of geographical latitude (**LAT**) and longitude (**LGT**), both in degrees.

It should be noted that time, which is so familiar to us in our daily lives, is expressed in different ways. The time by which we conduct our daily lives is called Central Standard Time (**CST**) or Greenwich time. **CST** for the United Kingdom is based on the meridian that passes through Greenwich, which is 0 degree longitude. The time for each particular place is calculated by its distance from this prime meridian. The earth is divided roughly into 24 time zones, by meridians 15 degrees apart (24x15=360). Japan is all in one time zone, and the **CST** for Japan is based on the longitude at Akashi city, 135 degrees. There is a difference between **CST** and solar time based on the position of the sun in the sky, except at the place where the longitude is adopted for **CST**; in Japan this place is Akashi. Even in places such as Akashi, there is some time difference between **CST** and the true solar time (**TST**) because the above calculation is based on the assumption that one rotation of the earth takes exactly 24 hours. This difference is called the equation of time (**EQT**). If we calculate mean solar time (**MST**) from **CST** and the difference in longitude, we have

$$\textbf{TST} = \textbf{MST} + \textbf{EQT} \qquad (5.14)$$

Using the relationship that one hour is equal to 15 degrees, it is convenient to express time in degrees and to call it the hour angle (**HAG**). The hour angle can be calculated as follows:

$$\textbf{HAG} = (\textbf{CST} + ((\textbf{ LGT} - \textbf{LGTstd })/\textbf{15}) + \textbf{EQT}) * \textbf{15} \qquad (5.15)$$

where **HAG** at solar time noon is zero and **LGTstd** is the longitude of **CST** for a particular region and for Japan is 135. **EQT** (in minutes) can be calculated using the following equation:

$$\textbf{EQT} = 9.87 * \textbf{sin}(2*\textbf{B}) - 7.53 * \textbf{cos}(\textbf{B}) - 1.5 * \textbf{sin}(\textbf{B}) \qquad (5.16)$$

where, **B** equals 2*pi* (**n**-81) / 364 and **n** is Julian day.
The sun's declination (**DEC**) is the angle between a line connecting the centers of the sun and the earth and the projection of this line on the earth's equatorial plane. **DEC** (in degrees) can be calculated using the following equation:

$$\textbf{DEC} = 23.45 * \textbf{sin} (((284+\textbf{n}) / 365)*(2*\textbf{pi})) \qquad (5.17)$$

On the vernal equinox (March 21, n=81) and autumnal equinox (September 21, n=264), **DEC** equals 0; on the summer solstice (June 21, **n**=172), **DEC** equals 23.45°; and on the winter solstice (December 21, **n**=355), **DEC** equals –23.45°.
Equations to calculate the sun's altitude and azimuth from the south for a horizontal surface are as follows:

$$\textbf{sin(SALT)} = \textbf{sin(LAT)}*\textbf{sin(DEC)} + \textbf{cos(LAT)}*\textbf{cos(DEC)}*\textbf{cos(HAG)} \quad (5.18)$$

$$\cos(SAZM) = (\sin(SALT)*\sin(LAT) - \sin(DEC))/\cos(SALT)/\cos(LAT) \quad (5.19)$$

The sun's declination and the equation of time for the year 1986 are given in Table 5.2; they do not change much from one year to another. If more exact values are needed, they can be obtained from an almanac or other reference.

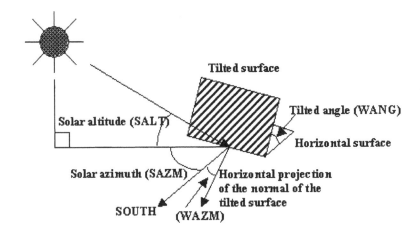

Figure 5.7. Solar angles for horizontal and tilted surfaces.

Table 5. 2. The sun's declination and the equation of time for the year 1986.

Day	1		8		15		22	
Month	DEC Deg:Min	Eq.of.Time min:s	DEC Deg:Min	Eq.of.Time min:s	DEC Deg:Min	Eq.of.Time min:s	DEC Deg:Min	Eq.of.Time min:s
January	-23:03	-3:16	-22:19	-6:26	-21:13	-9:13	-19:48	-11:20
February	-17:17	-13:31	115:09	-14:10	-12:51	-14:12	-10:23	-13:37
March	- 7:47	-12:30	15:05	-10:59	-2:20	-9:19	0:25	-7:07
April	4:20	-4:05	7:01	-2:04	9:35	-0:13	12:01	1:21
May	14:55	2:50	16:57	3:30	18:45	3:41	20:17	3:26
June	21:59	2:20	22:48	1:09	23:17	-0:15	23:27	-1:46
July	23:09	-3:38	22:32	-4:54	21:37	-5:50	20:23	-6:22
August	18:09	-6:19	16:17	-5:42	14:13	-4:35	11:57	-3:02
September	8:28	-0:12	5:53	2:05	3:14	4:32	0:31	7:01
October	-2:59	10:06	-5:51	12:14	-8:19	14:02	-10:52	15:14
November	-14:15	16:22	-16:25	16:16	-18:21	15:30	-20:01	14:03
December	-21:43	11:13	-22:33	8:24	-23:14	5:11	-23:27	1:46

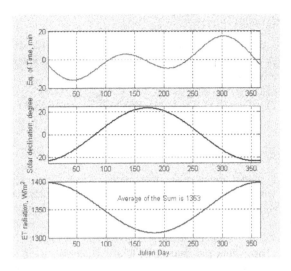

Figure 5.8. Output Figure Window of **cuc04b.m**.

The program, listed in Fig. 5.9, is provided to plot curves of equation of time (**EQT**, in min), solar declination (**DEC**, in degrees) and the daily solar constant as extraterrestrial (**ET**) radiation (W/m^2) throughout the year. The output is given as three output plots in one Figure Window as shown in Fig. 5.8.

```
% Program to plot equation of time, solar declination and        cuc04b.m
%          extraterrestrial radiation throughout the year.
h1=findobj('tag','cuc04b');close(h1);
figure('tag','cuc04b','Resize','on','MenuBar','none',...
   'Name','EQT','NumberTitle','off','Position',[120,60,520,480]);
x=1:1:365;          subplot(3,1,1);
B=2*pi*(x-81)/364;  EQT=9.87*sin(2*B)-7.53*cos(B)-1.5*sin(B);
plot(x,EQT,'r-','linewidth',2);
ylabel('Eq. of Time, min');axis([1 365 -20 20]);   grid on;
subplot(3,1,2);
dec=23.45*sin(2*pi*(284+x)/365);
plot(x,dec,'k-','linewidth',2);
ylabel('Solar declination, degree');axis([1 365 -25 25]);   grid on;
subplot(3,1,3);
Jsc=1353;       JOW=Jsc*(1+0.033*cos(2*pi*x/365));
plot(x,JOW,'b-','linewidth',2);          xlabel('Julian Day');
ylabel('ET radiation, W/m^2');  % ET is extraterrestrial
axis([1 365 1300 1400]);   grid on;
Solarconstant=sum(JOW)/365
tit=['Average of the Sum is ' num2str(Solarconstant)]
text(120,1370,tit);
```

*Fig. 5.9. Program to plot equation of time, solar declination and extraterrestrial radiation throughout the year (*cuc04b.m*).*

5.6.2. On a tilted surface

Suppose we have a tilted surface whose tilt angle is **WANG** (deg) and azimuth is **WAZM** (deg), as shown in Fig. 5.7; then the sun's altitude for the tilted surface (**SALTT**) is calculated by the following equation:

$$\sin(\text{SALTT}) \; = \; \sin(\text{SALT}) \; * \; \cos(\text{WANG}) \; + \; \cos(\text{SALT}) \; *$$
$$\sin(\text{WANG}) * \cos(\text{SAZM - WAZM}) \tag{5.20}$$

5.6.3. Transmitted solar radiation (CUC05)

Now consider a model to calculate transmitted direct solar radiation under a horizontal film when the time of the day is given. We already have a model to calculate transmitted solar radiation through a film, if the incident angle to the film (**THET**) is given (see Fig. 5.4). Use eq. 5.18 to find the sun's altitude. Then the incident angle of direct solar radiation at a given time of the day is 90 minus the sun's altitude. The whole program (**CUC05.m**) is given in Fig. 5.10a and two functions are listed in Fig. 5.10b (**RADcal.m**) and Fig. 5.4b (**RFcal.m**).

The program is dependent on time, but the relationships are entirely steady-state and are expressed sequentially. The material used in the simulation is a 1 mm thick glass sheet. The location is near Tokyo as defined by the latitude (**LATD**) and longitude (**LGT**). The value of **LGTstd** should be changed accordingly with the value of **LGT**. For example, **LGT** for New York City is around -121.5, so the **LGTstd** should be -120.

The same variable name can be placed on both sides of the equal sign in one equation, such as **SAZM=SAZM/CONV**, which gives the sun's azimuth in degrees converted from radian. The program consists of two parts. The first part is used to calculate the sun's altitude and then direct and diffuse radiation at given times; the second part is used to calculate transmissivity, absorptivity and reflectivity at the same times. The amounts of solar radiation transmitted through a film and absorbed by the film can be calculated using this program.

Several expressions can be used to find **SAZM**. In the present case, eq. 5.19 is used. In summer the solar azimuth changes from northeast to northwest through east, south and west. If the solar azimuth is taken beginning at the south, it decreases from over 90 degrees to zero and then increases to over 90 degrees. Although the expression in eq. 5.19 for **cos(SAZM)** can give these angles for **SAZM** correctly, that in eq. 5.18 for **sin(SAMZ)** can give only the angle between 0 and 90 degrees because of restriction by **asin**.

```
% Program to calculate sun's position and transmitted          CUC05.m
% direct solar radiation through a horizontal film sheet
% Functions required: RFcal.m, RADcal.m
%
clear all; clc;
FL = 1;                          % FL: Thickness (mm)
FN = 1.526;                      % FN: Index of refraction
FK = 0.0441;                     % FK: extinction coefficient (1/mm)
```

```
CONV = pi/180;                    % CONV: Conversion factor
JW = 1360;                        % JW: Solar constant (W/m2)
J0 = JW*3.6;                      % J0: Solar constant (kJ/m2/hr)
pp = 0.7;                         % pp: atm. transmittance (ND)
%------ Data for Tokyo area ------------------------------------
LATD = 35.68;                     % LATD: Latitude, 35 deg 41 min
LAT = LATD*CONV;                  % LAT:  in Radiant
LGT = 139.77;                     % LGT:  Longitude, 139 deg 46 min
LGTstd=135;                       % LGTstd: std. longitude of Japan
%------ Data for October 15 from Table 5.2 ---------------------
DEC = -8.32;                      % DEC: Declination, -8 deg 19 min
DEG = DEC*CONV;                   % in Radiant
EQT = 0.234;                      % EQT: eq.of time,14 min 02 sec
%
t24=48;
TRAN=zeros(1,t24); REFL=zeros(1,t24); ABSO=zeros(1,t24);
RAD=zeros(1,t24); RADS=zeros(1,t24);
WRAD=zeros(1,t24); WRADS=zeros(1,t24);
IRAD=zeros(1,t24); IRADS=zeros(1,t24);
SALT=zeros(1,t24); SAZM=zeros(1,t24);
for t = 1:48                      % for 24 hr
  CST = t/2 - 12;
  % Central standard time is taken from noon and is positive in the afternoon
  HAG = (CST+((LGT-LGTstd)/15)+EQT)*15;     % Hour Angle, in degree
  HAG = HAG*CONV;                           % in Radiant
  SALT(t)= asin(sin(LAT)*sin(DEG)+cos(LAT)*cos(DEG)*cos(HAG));
  SAZM(t)=acos((sin(SALT(t))*sin(LAT)-sin(DEG))/cos(SALT(t))/cos(LAT));
  % SALT and SAZM both in Radiant
  if (SALT(t) <= 0)
    TRAN(t)=0; REFL(t)=1; ABSO(t)=0; RAD(t)=0; RADS(t)=0;
    SALT(t)=0; SAZM(t)=999;
  else
    % Calculation of solar radiation
    [RAD(t),RADS(t)]=RADcal(J0,SALT(t),pp); % Calling function RADcal()
    THET = pi/2-SALT(t);                    % in Radiant
    SALT(t)= SALT(t)/CONV;                  % in Degree
    SAZM(t)= SAZM(t)/CONV;                  % in Degree
    [RF, CTHETP]=RFcal(THET,FN);            % Calling function RFcal()
    AB=exp(-FK*FL/CTHETP);
    TRAN(t)=(1-RF)*(1-RF)*AB/(1-RF*RF*AB*AB);
    REFL(t) = RF+RF*(1-RF)*(1-RF)*AB*AB/(1-RF*RF*AB*AB);
    ABSO(t) = 1-RF-(1-RF)*(1-RF)*AB/(1-RF*AB);
  end
  WRAD(t) = RAD(t)/3.6; IRAD(t)=WRAD(t)*TRAN(t);
  WRADS(t) = RADS(t)/3.6;IRADS(t)=WRADS(t)*TRAN(t);
  % Assuming transmittance of direct and diffuse sunlight are the same
end
%[Figure 1]----------------------------------------------------
x=1:48;
h1=findobj('tag','cuc05_part1'); close(h1);
figure('tag','cuc05_part1','Resize','on','MenuBar','none',...
         'Name','CUC05.m (Figure 1: Solar Angles vs. Time)',...
         'NumberTitle','off','Position',[140,80,520,420]);
plot(x/2, SALT,'r+-', x/2,SAZM,'b*-', 'linewidth',2);
xlabel('Time, hour'); ylabel('Solar Angles, degree');
legend('Solar altitude','Solar azimuth',0);
axis([1,24,-inf,90]); grid on;
%[Figure 2]----------------------------------------------------
h1=findobj('tag','cuc05_part2'); close(h1);
figure('tag','cuc05_part2','Resize','on','MenuBar','none',...
```

```
           'Name', 'CUC05.m (Figure 2: Solar Radiation vs. Time)',...
           'NumberTitle','off','Position',[180,60,520,420]);
plot(x/2,WRAD,'r+-', x/2, WRADS,'b*-', ...
    x/2, IRAD,'r^-.',x/2,IRADS,'bv-.','linewidth',2);
xlabel('Time, hour'); ylabel('Solar Radiation, W/m^2');
legend('Direct (Outdoor)','Diffuse (Outdoor)',...
    'Direct (under glazing)','Diffuse (under glazing)');
axis([1,24,-inf,inf]); grid on;
%[Figur 3]------------------------------------------------
h1=findobj('tag','cuc05_part3'); close(h1);
figure('tag','cuc05_part3','Resize','on','MenuBar','none',...
           'Name', 'CUC05.m (Figure 3: Optical properties vs. time)',...
           'NumberTitle','off','Position',[220,40,520,420]);
subplot(1,1,1);
plot(x/2,TRAN,'r+-', x/2, REFL,'bo-', x/2, ABSO,'k*-','linewidth',2);
axis([1,24,0,1]); xlabel('Time, hour');
ylabel('Tran., Refl. Abso. (ND)');
legend('Transmissivity','Reflectivity','Absorptivity',0); grid on;
clc; disp('You can enter ''close all'' to close figure windows.');
```

*Figure 5.10a. Program to calculate time courses of solar angles, solar radiation and optical properties of glazing (**CUC05.m**).*

In the present program, calculation of azimuth is not included because it is not necessary. If azimuth is included, some consideration should be given to computer error because the time step in these calculations is small. If the calculation begins at exact solar noon -- that is, azimuth 0 -- acos(0) should be 1, but it can be slightly larger than 1. An error message told us this had happened in our early calculations. The discrepancy can be solved by filtering, such as we have in the program in Fig. 5.10b. We have an **if** statement that holds **ALTS** at not less than 0.01. If you look at the remaining several statements, it is not difficult to understand why it is necessary to have this filter. The value of **PP** is less than 1; therefore it is meaningless to calculate **PPP** with a large value of **ALTC**.

```
% RADcal.m
function [RAD,RADS]=RADcal(J0,SALT,pp)
ALTS=sin(SALT);
if ALTS < 0.01,ALTS=0.01;end  % filtering
ALTC=1/ALTS; ppp=pp^ALTC; RAD=J0*ALTS*ppp;
RADS=J0*ALTS*(1-ppp)/(1-1.4*log(pp))/2;
```

*Figure 5.10b. Function required in program **CUC05.m** (**RADcal.m**).*

The results of program 'CUC05.m' are shown in Fig. 5.11. In total, three figures were drawn. Fig. 5.11a shows the time courses of solar angles. The times of sunrise and sunset are quite clear from this figure. The time for sunrise is the solar altitude changes from zero to a positive value, and the time for sunset is when the altitude returns to zero. Also, at solar noon, the solar azimuth is zero, and the solar altitude is at its peak.

Fig. 5.11b shows the time courses of direct and diffuse solar radiations outside and under the glazing, assuming the transmittance of direct and diffuse sunlight are the same.   In the early morning and late afternoon, of the components of total solar radiation, the diffuse sunlight contributes more than the direct sunlight.

Fig. 5.11c shows the time courses of optical properties of glazing.   During the dark period (from 6 pm to 6 am), the reflectance shows a value of 1.   This should not be interpreted as 100% reflection of sunlight for there is no sunlight at all during the dark period.   It is simply the result of **TRAN** + **REFL** + **ABSO** =1.   During the dark period, the **ABSO** and **TRAN** are zero, thus leading to **REFL** = 1.

It is possible that the legend can block part of the curves.   In such situations, users can use the mouse to click on the legend and drag it to a better location as shown in Fig. 5.11c.

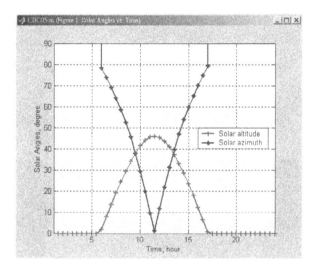

Figure 5.11a. Time courses of solar angles (Figure 1 generated by **CUC05.m***).*

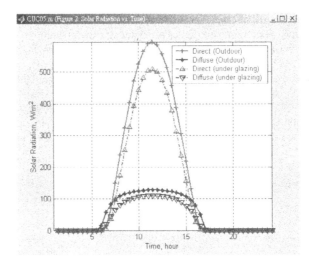

Figure 5.11b. Time courses of solar radiation above and under glazing (Figure 2 generated by **CUC05.m***).*

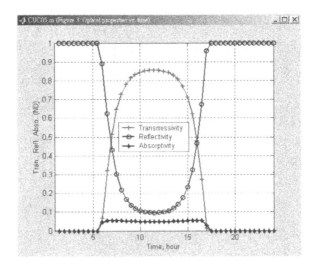

Figure 5.11c. Time courses of optical properties of a film (Figure 3 generated by CUC05.m).

PROBLEMS

1. Rerun the program **CUC04**, applying different values for **FN** and **FK** as shown in Table 5.1.

2. The thickness of the glass (**FL**) is assumed 3 mm in the model **CUC04**. Change it to 1 and 5 mm and generate the transmissivity curves.

3. Choose a day of the year other than Oct. 15 and use the data from Table 5.2 to rerun the program **CUC05**.

4. Calculate the hour angle at your location by using eq. 5.14 .

5. In the model **CUC05**, the longitude of the standard time location is fixed to 135 degrees. Change it to your standard time.   For U.S.A., **EST** is -75, **CST** is -90, **MST** is -105, and **PST** is -120, respectively.   Then input the latitude and longitude of your location as **LATD** and **LGT** and run the program.   The date is set for October 15.   You can also select other days using Table 5.2.

6. Write a program to derive the direct transmittance for materials listed in Table 5.1 when incident angle equals 0. Note: make use of '**RFcal.m**'.

7. In the model of **CUC05**, **DEC** and **EQT** are selected from Table 5.2. Use eqs. 5.16 and 5.17 to calculate **DEC** and **EQT** with given Julian day (n=288 for October 15) and rerun the model.

CHAPTER 6

TEMPERATURE ENVIRONMENT UNDER COVER

6.1. INTRODUCTION

Temperature is another important environmental condition for plant growth. The temperature of the plant environment can be modified by cover the area. In the daytime, the main energy source is, of course, the sun, and its energy is stored in the soil. The transmissivity of the film used for covering is an important factor in getting a large amount of energy into the soil surface. At night, the energy source is thus the soil layer. The main heat loss from the surface is due to long wave radiation, and the emissivity of the cover is a key factor in this.

Long wave radiation heat exchange between the sky and the soil surface is expressed by the difference between eqs. 4.4 and 4.17, where the absolute temperature of the soil surface is **AT**. Emissivities for covering materials range between 0 and 1; they are shown in Table 2.3 in Chapter 2, where emissivity is expressed as absorptivity.

6.2. EFFECT OF MULCHING

6.2.1. Experimental measurements

The effect of mulching is well appreciated, but its mechanism is not yet fully understood. Many field experiments have been conducted over a period of years. But it is rather difficult to isolate boundary conditions in field experiments, so the effect of mulching on the temperature in the surrounding environment has not been clearly explained. For example, suppose we set up three mulchings with three films with different color -- clear, black and white -- in order to examine the temperature differences in the soil layer. Even if the soil condition appears to be the same under the three coverings, the actual situation will be different. There will be slight differences in the way the coverings are placed. Air leakage through the covering affects water vapor conditions under the cover, and therefore the temperature. Measurement of soil temperature is not easy: the spacing and the placement of the sensing element at a certain point requires considerable skill. Even if thermocouples are used, they are not easy to place exactly 1 cm deep in the soil. These are all reasons to use computer simulation.

Figure 6.1 shows one of the typical results of soil temperature measurement under different mulches, and Table 6.1 lists some major components of the heat balance in these systems. These are all measured values. Soil temperatures 1 cm deep under mulchings of various materials are valuable not only as good indicators for comparison but also determining the variability of the measurements. Since details on the covering materials are not shown, quantitative comparisons cannot be made on the basis of these data. It is clear from Table 6.1 that there is some difference in latent heat transfer among these coverings. However, we cannot quantify it because of differences in the permeability of these materials and because of measurement errors. Furthermore, vapor transfer through the holes made for plants might affect the situation. In Table 6.1, albedo (reflectivity of solar radiation) is given and surface temperature is not. The techniques for making simulation models in the present chapter show that surface temperatures are easily derived and that transmissivity and absorptivity rather than albedo are needed to analyze the energy balance of these systems.

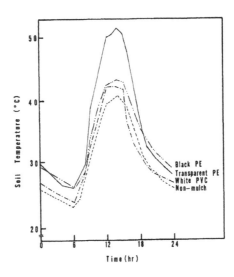

Figure 6.1. Daily variation of soil temperature at 1 cm depth under various types of mulching in Aug., 1984 (after Kwon, 1988).

Table 6.1. Some major components of heat balance under various mulchings (after Kwon, 1988).

Treatment	Albedo (%)	Net radiation	Sensible heat	Latent heat	Soil heat flux	Solar radiation	**Volumetric heat capacity
Non-mulched	11.8	905.5	510.3	274.7	120.5 (100)	2368	1.76
Transparent PE	19.4	889.1	685.9	55.4	147.8 (123)	2368	2.10
White PVC	32.1	1100.4	978.2	37.4	84.8 (71)	2368	2.06
Black PE	4.3	779.1	621.6	45.8	111.7 (93)	2368	2.18
Straw mulch	21.4	1300.3	1077.7	136.9	85.7 (71)	2368	2.14

All units except volumetric heat capacity are $Jcm^{-2}day^{-1}$ and were investigated on Aug. 19, 1984.
** Unit is $Jcm^{-3o}C^{-1}$ and was investigated at soil depth between 10 and 20 cm on Oct. 11, 1984.

6.2.2. A model to simulate temperature regime under mulching (CUC20)

A computer simulation enables us to isolate boundary conditions and idealize the situation.   Let us look at a model that compares the three colors of mulching: clear, black, and white.   The main part of the model is depicted in Fig. 6.2.   This part is the surface of the soil layer where film mulching has been used.   The air gap between the film and the soil surface is enlarged in order to show the energy exchange in this layer clearly.

It is assumed that the air temperature in this layer is the arithmetic mean of the film temperature (**TC**) and the soil surface layer temperature (**TF**) and that the air layer is at the same temperature as the surface of the soil which is saturated with water vapor.   All energy terms involved in the energy balance of the film to determine its temperature are shown in the Figure.   They include direct solar radiation (**RAD**), diffuse radiation (**RADS**), long wave radiation from both sides of the film surface (**EPSC*SIG*TC4**), thermal radiation from the soil surface (**EPSF*SIG*TF4**) and from the sky (**EPSA*SIG*TO4**), convective heat transfer from both sides of the film **HO*(TC-TO)** and **HI*(TC-TI)**, and energy-related condensation of water vapor at the inside surface (**HWT**).   Evaporation from the soil surface is expressed in a way similar to that in Chapter 4, but a factor (0.0 - 1.0) to reduce the saturated humidity ratio (Greenhouse Soil Index; GSI) is introduced. The thermal radiation from the film to the soil surface and then reflected back from the soil surface is also considered.

The program of the model for mulching is shown in Figs. 6.3a, b and Figs. 6.4a, b and the simulation results are given in Figs. 6.5, 6.6 and 6.7.   Enter 'cuc20(n)' in the Command Window of **MATLAB** to run this model.   Within the parentheses, n = 1 is for selecting clear mulch, n = 2 for black and n = 3 for white mulch.   The user can run these options one by one or enter 'cuc20(1); cuc20(2); cuc20(3)' followed by the Enter to run three options in a row.   Entering 'cuc20' along will assume n = 1.   This part of the program was done by checking the number of arguments (**nargin** is a reserved word of **MATLAB**).   If no argument exists, the program will assign 1 to the parameter **out**.

Fig. 6.3a shows the branching function **switch…case ….end** both before and after the line containing the **ode23t** function.   The former **switch** is used to choose the type of film and the latter part is to draw the corresponding Figure of the chosen film.   A stiff solver is suggested when solving simultaneous ordinary differential equations (ode) of this model.   Function **ode23t** provides the shortest execution time.

Fig. 6.3b shows the 'soil20.m' subprogram containing all the simultaneous equations related to the energy balance of the cover, inside air layer, soil surface and soil layers.   The main part of the model is depicted in Fig. 6.2.

In the present model, function **FRSNL** (listed in Fig. 6.4a) is defined to calculate transmissivity and absorptivity of the film and direct and diffused solar radiation for clear PE film, and function **FRSNLa** (listed in Fig. 6.4b) is for black and white PE films.

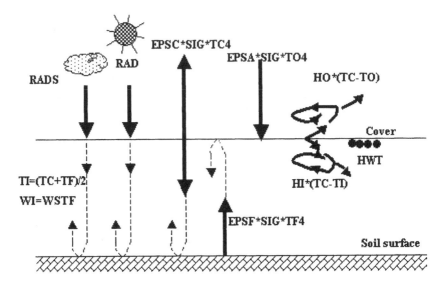

Figure 6.2. Energy balance of mulching.

Calculated coefficients **TDS** and **TLV** are transmissivity of the film for diffuse solar radiation and long-wave radiation, respectively. The amount of condensation at the inside surface of the film is calculated in the '% **inside air layer**' section as shown in Fig. 6.3b. In this section, the saturated humidity ratio at the inside air temperature is compared with that at the film temperature. The amount of condensed water is **TW**, and its latent heat transfer is **HWT**.

The soil surface is assumed sealed by a thin film. No air leakage from air space between the film and soil surface is assumed, and therefore the air gap is saturated by water vapor. Heat transfer in this air gap is also assumed to take place by conduction only.

The thickness of the covering film (**FL**) is assumed to be 1 mm and is defined as a **global** variable in this program as listed in subprograms shown in Figs. 6.3b, 6.4a and 6.4b. Normal thickness is about one tenth of the thickness used in this program. The reason for this change is simulation stability. If the film is thinner than 1 mm, it is difficult to obtain a stable solution by numerical iteration. On the other hand, it is clear that the resistance at both film surfaces mostly governs heat transfer through thin film and the effect of the difference in film thickness can be neglected if the film is less than several millimeters thick. This can be verified by the fact that the thickness of normal glass for greenhouses is 3-4 mm and there is no difference between glasshouses and plastic film houses attributable to the difference in covering thickness.

```
%  Temperature regime in soil under CLEAR/BLACK/WHITE          CUC20.m
%       PE Mulching. Latent heat transfer is involved
%  function required: soil20.m
function cuc20(out)
global colr;
if nargin==0, out=1;end
switch out
case 1
   clc
   colr='clear';opt='Clear PE Mulch';
case 2
   colr='black';opt='Black PE Mulch';
case 3
   colr='white';opt='White PE Mulch';
end
disp(opt);
tic
t0=1; tfinal=48;
y0=[10; 10; 20; 20; 20; 20];
%   TC, TF, T1, T2, T3, T4    (Initial condition)
%
[t,y]=ode23t('soil20',[t0 tfinal],y0);
%
% Use Stiff Solver, time required to run
% ode23t < ode15s  < ode23tb < ode23s
% Use Non-stiff Solver
% ode23, ode45 and ode113 will take too long to solve
%
switch out
case 1
   h1=findobj('tag','cuc20_part1');   close(h1);
   figure('tag','cuc20_part1','Resize','on','MenuBar','none',...
      'Name','CUC20.m (Figure 1)','NumberTitle','off',...
      'Position',[140,120,520,420]);
case 2
   h1=findobj('tag','cuc20_part2');   close(h1);
   figure('tag','cuc20_part2','Resize','on','MenuBar','none',...
      'Name','CUC20.m (Figure 2)','NumberTitle','off',...
      'Position',[180,80,520,420]);
case 3
   h1=findobj('tag','cuc20_part3');   close(h1);
   figure('tag','cuc20_part3','Resize','on','MenuBar','none',...
      'Name','CUC20.m (Figure 3)','NumberTitle','off',...
      'Position',[220,40,520,420]);
end
plot(t,y(:,1),'kh-',t,y(:,2),'mo--',t,y(:,3),'rx-',t,y(:,4),...
   'c+-',t,y(:,5),'gs-',t,y(:,6),'b*-');
% 'kh-' : black,    hexagram, solid line for curve TC
% 'mo-' : magenta, circle,   solid line for curve TF
% 'rx-' : red,      x-mark,   solid line for curve T1
% 'c+-' : cyan,     plus,     solid line for curve T2
% 'gs-' : green,    square,   solid line for curve T3
% 'b*-' : blue,     star      solid line for curve T4
grid on;
titcont=['Cover and soil temperatures under ' opt];
title(titcont);         xlabel('time elapsed, hr');
ylabel('Temperature, ^oC'); legend('TC','TF','T1','T2','T3','T4');
toc
% 'tic'&'toc' are pair of functions to measure run time.
%
```

```
if out==3
  disp('Thank you for using CUC20:');
  disp('      Temperatures under CLEAR/BALCK/WHITE PE mulch'); disp(' ');
end
```

Figure 6.3a. Program to calculate cover and soil temperatures under mulch (**CUC20.m**).

```
% Function to be used with CUC20.m                        soil20.m
% Also involves functions FRSNL.m, FRSNLa.m, FWS.m, TABS.m
%
function dy = soil20(t,y)
global FL;
global colr;
T0=15.0; TU=5.0; TBL=20.0;% Temp (C)
TD=8;                    %outside dew point temperature, in degree C
KS = 5.5; CS= 2.0E+3;
% KS (kJ/m/C/hr) and KS/3.6 (W/m/C) also CS (kJ/m3/C)
CA=1.164; CC=50.0;
RHO=1.164;
GSI=1.0;    % Greenhouse Soil Index, dryness of soil surface (01-1.0)
SIG= 20.4; % Stefan-Boltzmann constant (kJ/m2/K4/hr) = 5.67(W/m2/K4)
HI=7.2;    HO= 25.4;
Z0=0.01;   Z1=0.05;   Z2=0.1;   Z3=0.2;
Z4=0.5;    % Depths of soil layer (m)
ALC=0.1;   ALF=0.7;   RMC=0.1;
RMSC=0.05; EPSC=0.15; EPSF=0.95;
EPSA = 0.711+(TD/100)*(0.56+0.73*(TD/100));
HLG=2501.0; LE=0.9;
KM=3.6*HI/LE/CA/RHO;
FL=1;DEX=0.001*FL; %Thickness in mm and m
% EPSA:Emissivity of air layer
WO=FWS(TD);                     % calling function FWS()
TDS=1-ALC-RMSC;
TLV=1-EPSC-RMC;
%
clk = mod(t,24);
if colr=='clear'
   [TRAN,ABSO,RAD,RADS]=FRSNL(clk);   % calling function FRSNL()
else
    [TRAN,ABSO,RAD,RADS]=FRSNLa(clk); % calling function FRSNLa()
end
OMEGA=2*pi/24;                        % Time (hr)
TO = T0 + TU*sin(OMEGA*(clk-8));
%-------------------------------------------------------------------
TC=y(1); TF=y(2);T1=y(3);T2=y(4);T3=y(5);T4=y(6);
% At Cover ---------------------------------------------------------
TO4=TABS(TO); TC4=TABS(TC); TF4=TABS(TF); % calling function TABS()
% calculating HWT
HWT=0;
WSS=FWS(TC);              % Calling function FWS()
% Inside air layer ------------------------------------------------
TI=(TC+TF)/2;
WSTF=FWS(TF);            % calling function FWS()
WI=WSTF;
if WSS < WI
   TW=WI-WSS;           % amount of water condensed
   WI=WSS;
   HWT=TW*HLG*KM;       % Energy required to condense water
end
%
```

```
ITC=(RAD*ABSO+RADS*ALC+(1-ALF)*(RADS*TDS+RAD*TRAN)*ALC+...
   SIG*EPSC*(EPSF*TF4+ EPSA*TO4+((1-EPSF)-2)*TC4)-...
   HO*(TC-TO)-HI*(TC-TI)+HWT)/DEX/CC;
% At soil surface ------------------------------------------------------
ITF=(ALF*(RADS*TDS+RAD*TRAN)-KS*(TF-T1)*2/(Z0+Z1) ...
   -HI*(TF-TI)+HLG*KM*(WI-GSI*WSTF)...
   -SIG*EPSF*((1-RMC*EPSF)*TF4-EPSA*TLV*TO4-EPSC*TC4))/CS/Z0;
% In soil layers ------------------------------------------------------
IT1 = (KS*(TF - T1)*2.0/(Z0+Z1)+KS*(T2 - T1)*2.0/(Z1+Z2))/CS/Z1;
IT2 = (KS*(T1 - T2)*2.0/(Z1+Z2)+KS*(T3 - T2)*2.0/(Z2+Z3))/CS/Z2;
IT3 = (KS*(T2- T3)*2.0/(Z2+Z3)+ KS*(T4 - T3)*2.0/(Z3+Z4))/CS/Z3;
IT4 = (KS*(T3 - T4)*2.0/(Z3+Z4)+KS*(TBL - T4)*2.0/Z4)/CS/Z4;
dy=[ITC; ITF; IT1; IT2; IT3; IT4];
return
```

Figure 6.3b. One of the subprograms of CUC20 model (**soil20.m**).

6.2.3. Same model for black or white mulch

Transmissivity and absorptivity of clear film are calculated by the equations given in the preceding section, and source codes listed in function **FRSNL** can be found in Fig. 6.4a. For opaque black and white films transmissivity and absorptivity are given by interpolation functions based on experimental observations. Source codes listed in function **FRSNLa** can be found in Fig. 6.4b. As shown in the source code, type of mulching is assigned to the parameter **colr**, which decide the proper data set for the interpolation of the calculation of **TRAN** and **ABSO** in function **mulching**.

Film properties such as transmissivity and absorptivity of black and white opaque films are given in function **mulching(colr),** which is in the last part of **FRSNLa.m** (Fig. 6.4b).

```
% Function for Clear PE Mulching to be used in CUC20 model      FRSNL.m
% Also involves functions: RADcal.m, RFcal.m
%
function [TRAN, ABSO, RAD, RADS]=FRSNL(clk)
global FL pp ARTO
LATD=35.68;LGT=139.77; LGTstd=135;conv=pi/180;
% LATD: latitude 35 deg 41 min is f(location)
% LGT: longitude 139 deg 46 min is f(location)
% LGTstd: meridian 139 longitude is f(location)
LAT=LATD*conv; % latitude in radian unit
DEC=-8.316;EQT=0.2338;% data for October 15 from Table 5.1
DEG=DEC*conv;
% DEC: declination -8 deg 19 min is f(Julian day)
% EQT: Equation of time 14 min 2 sec is f(Julian day)
% DEG: declination
JW=1360;   % JW:  solar constant (W/m2)
J0=JW*3.6; % J0:  solar constant (kJ/m2/hr)
pp=0.7;    % pp:  atmospheric transmittance
FN=1.526;FK=0.0441;
% FN: Index of reflection
% FK: extinction coefficient (1/mm)
CST=clk-12;
HAG=(CST+((LGT-LGTstd)/15)+EQT)*15;  HAG=HAG*conv;
SALT=asin(sin(LAT)*sin(DEG)+cos(LAT)*cos(DEG)*cos(HAG));
if SALT<=0
   TRAN=0;REFL=1;ABSO=0;RAD=0;RADS=0;
```

```
else
   [RAD, RADS]=RADcal(J0,SALT,pp);        % Calling function RADcal()
   THET=pi/2-SALT;
   [RF, CTHETP]=RFcal(THET,FN);           % Calling function RFcal()
   AB=exp(-FK*FL/CTHETP);
   TRAN=(1-RF)*(1-RF)*AB/(1-RF*RF*AB*AB);
   REFL=RF+RF*(1-RF)*(1-RF)*AB*AB/(1-RF*RF*AB*AB);
   ABSO=1-RF-(1-RF)*(1-RF)*AB/(1-RF*AB);
end
return
```

Figure 6.4a. Subprogram to calculate transmissivity and absorptivity of clear film
(FRSNL.m).

The one-dimensional interpolation function **Interp1** is a very convenient tool for generating a non-linear relationship based on observed values in the following way:

$$\mathbf{YI = INTERP1(X, Y, XI, 'method')} \qquad (6.1)$$

Available methods are:

'nearest'	- nearest neighbor interpolation
'linear'	- linear interpolation
'spline'	- cubic spline interpolation
'cubic'	- cubic interpolation

The default method is linear interpolation. **X** and **Y** are vectors containing x and y values, respectively. **XI** is the value of **X** of interests and **YI** is the derived **Y** value corresponding to **XI**. All the interpolation methods require that **X** be monotonic. **X** can be non-uniformly spaced. For faster interpolation when **X** is equally spaced and monotonic, use the methods '*linear', '*cubic' or '*nearest'.

```
% Functions for Black and White PE Mulch in CUC20 model      FRSNLa.m
% Also involved function(s): RADcal.m, RFcal.m
%
function [TRAN, ABSO, RAD, RADS]=FRSNLa(clk)
global FL pp GSI;
global colr;
conv=pi/180;
JW=1360;    J0=JW*3.6;      pp=0.7;
LATD=35.68; LAT=LATD*conv; LGT=139.77;      LGTstd=135;
DEC=-8.32;  DEG=DEC*conv;  EQT=0.234;
FN=1.526;   FK=0.0441;
CST=clk-12;
HAG=(CST+((LGT-LGTstd)/15)+EQT)*15;         HAG=HAG*conv;
SALT=asin(sin(LAT)*sin(DEG)+cos(LAT)*cos(DEG)*cos(HAG));
if SALT<=0
   TRAN=0;REFL=1;ABSO=0;RAD=0;RADS=0;
else
   [RAD, RADS]=RADcal(J0,SALT,pp);% Calling function RADcal()
   THET=pi/2-SALT;
   ang=THET/conv;
   [Xtran, Ytran, Xabso, Yabso]=mulching(colr);
   %colr='black' or colr='white'; colr is a global variable
```

```
TRAN=interp1(Xtran,Ytran,ang,'nearest');
ABSO=interp1(Xabso,Yabso,ang,'nearest');   % Interpolation method 1
%TRAN=interp1(Xtran,Ytran,ang,'linear');
%ABSO=interp1(Xabso,Yabso,ang,'linear');   % Interpolation method 2
% method 3 is 'spline' and method 4 is 'cubic'
% # of pts is not enough in this example for methods 3 & 4.
end
return
%-----------------------------------------------------------------------
function [Xtran, Ytran, Xabso, Yabso] = mulching(colr)
switch colr
   case 'black'        % Black PE Mulch
      Xtran=[0 10 60 65 80 90];   Ytran=[0.05 0.05 0.05 0.045 0.03 0.0];
      Xabso=[0 10 88 90];         Yabso=[0.9 0.9 0.9 0.0];
   case 'white'        % white PE Mulch
      Xtran=[0 10 60 65 80 90];   Ytran=[0.58 0.58 0.58 0.5 0.4 0.0];
      Xabso=[0 10 88 90];         Yabso=[0.05 0.05 0.05 0.0];
end
return
```

Figure 6.4b. Subprogram to calculate transmissivity and absorptivity of black and white opaque films (**FRSNLa.m**).

6.2.4. Results of CUC20 model

In Fig. 6.5, the temperature patterns of each component are shown.   In general, soil temperatures with mulching (model **CUC20**) are much higher than those of soil without mulching (model **CUC03**), mostly because mulching prevents cooling of the soil surface due to evaporation.   In Fig. 6.5, Comparison of maximum values also shows that the temperature of the soil surface (**TF**) is the highest because of the transmissivity of the covering film.   It is assumed that the clear film is just 1 mm thick and that its transmissivity is very high.   The soil surface is therefore heated by the large amount of transmitted solar radiation.   Because of their limited heat capacities, the temperatures of these components drop when they are not heated. Among the components in the model, the film (**TC**) has the lowest minimum temperature.   The components are cooled by way of radiative cooling.

Under which mulching is the soil temperature highest: clear, black, or white? Let us compare the simulation results in Figs. 6.5, 6.6, and 6.7.   First of all, the film temperature is the highest for black mulch mainly because of its radiation absorptivity.   There is some variation in the minimum temperatures of films, but the difference is small and can be considered negligible.

Maximum soil surface temperature is the highest for clear mulch, followed by white and black, in that order.   It is clear that the main factor for this is transmissivity.   In the present calculations, it is assumed that transmissivity is 0.85 for clear, 0.58 for white, and 0.05 for black film at $0°$ incident angle.   The difference between black and white films is caused by the difference in their absorptivity, which is assumed to be 0.05 for white film and 0.9 for black film. White mulching reflects more radiation back to the air, while black mulching absorbs it.   Both reflected and absorbed radiation are used to heat the film and to some extent warm the soil.   This tendency governs all soil layers.   Minimum temperatures or night time temperatures are almost the same for these mulchings but are highest by a small margin for clear film, followed by white and black, in the same order as the maximum temperatures in the daytime.

It should be noted that these soil temperature regimes under mulching are governed by film properties such as absorptivity, transmissivity and reflectivity. Typical values for each film are used here, but measured values for your own use are highly recommended if you need exact comparisons.   Outside input conditions such as solar radiation and dew point temperature are also adjustable for various experimental conditions.

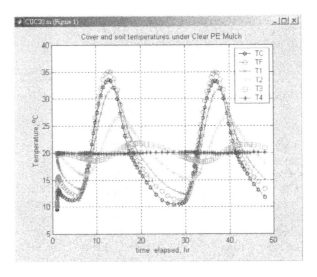

Figure 6.5. Cover and soil temperatures under clear PE mulch, result of cuc20(1).

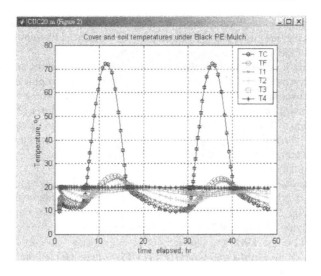

Figure 6.6. Cover and soil temperatures under black PE mulch, result of cuc20(2)

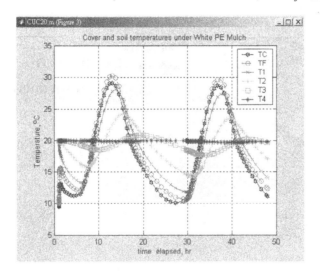

Figure 6.7. Cover and soil temperatures under white PE mulch, result of cuc20(3).

6.3. TEMPERATURE ENVIRONMENT UNDER ROW COVERS (**CUC30**)

Inflating the air gap between the cover and the soil surface in mulching can recreate the environment of row covers. Row covers are not flat; therefore, irradiation is different from mulching in its details. But the overall nature could be simulated using a simple model neglecting this difference. The only difference is that in the air layer, between the cover and the soil surface, air is replaced by ventilation. The heat balance of the inside air is depicted in Fig. 6.8.

In mulching it is assumed that the air between the cover and the soil surface is saturated with water vapor and that the air temperature is the arithmetic average of the soil and cover surface temperatures. In the present model, the air temperature between these two surfaces is one of the unknown variables and must be defined in an integral form. The heat balance of inside air is rather easy to calculate because of the lack of thermal radiation terms. Energy comes through the cover, soil, and ventilation (see Fig. 6.8). Latent heat transfer also occurs at the soil surface and is caused by ventilation.

The whole model is shown in Figs. 6.9a and 6.9b, and the computed result is given in Figs. 6.10a, and 6.10b. Closely look at the **% Inside air layer** part of the program in Fig. 6.9b. **ITI** is net energy gain and **TI** is inside air temperature. **IWI** is net mass gain and **WI** is inside humidity ratio. All components in the equation of **ITI** and **IWI** are shown in Fig. 6.8: energy gain from the cover, **HI*(TC-TI)**, from ventilation, **CA*QH*(TO-TI)** where **QH** is the airflow rate $(m^3/hr/m^2)$, and from the soil surface, **HI*(TF-TI)**. Also, mass gain due to ventilation, **RHO*QH*(WO-WI)** and that from the soil surface, **KM*(GSI*WSTF-WI)**.

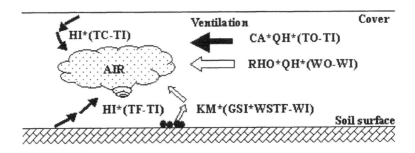

Figure 6.8. Energy balance of inside air.

```
% Model for Rrow Cover                                CUC30.m
% function involves: soil30.m
%
clear all;clc
global FL pp GSI QH
fprintf('\nPlease wait.\n');
t0=1;        tfinal=48;
y0=[10; 10; 0.01; 10; 20; 20; 20; 20];
%   TC, TI, WI,   TF, T1, T2, T3, T4  (initial condition)
%
[t,y]=ode23t('soil30',[t0 tfinal],y0);
%
h1=findobj('tag','cuc30_part1');        close(h1);
figure('tag','cuc30_part1','Resize','on','MenuBar','none',...
    'Name','CUC30.m (Figure 1)',...
```

```
'NumberTitle','off','Position',[120,120,520,420]);
plot (t,y(:,1),'r:',t,y(:,2),'b-.',t,y(:,4),'k-',t,y(:,5),'m--');
grid on;
axis([-inf, inf, 5, 40]);
titcont=['Temperatures changes under Row tunnel, given pp='...
    num2str(pp,3),', GSI=' num2str(GSI,3),', QH=' num2str(QH,4)];
title(titcont);
xlabel('time elapsed, hr'); ylabel('Temperature, ^oC');
legend('T_c_o_v_e_r','T_i_n_s_i_d_e_ _a_i_r','T_f_l_o_o_r','T_1');
h1=findobj('tag','cuc30_part2'); close(h1);
figure('tag','cuc30_part2','Resize','on','MenuBar','none',...
    'Name','CUC30.m (Figure 2)',...
    'NumberTitle','off','Position',[160,80,520,420]);
plot (t,y(:,3)); grid on;
axis([-inf, inf, 0, inf]);
xlabel('time elapsed, hr');
ylabel('Humidity Ratio, kg vapor/kg dry air');
titcont=['HR changes inside air layer of Row tunnel, given pp='...
    num2str(pp,3),', GSI=' num2str(GSI,3),', ...
    QH=' num2str(QH,4)];
title(titcont);
clc
  disp('Thank you for using CUC30');
  disp('');
  disp('You can enter ''close all'' to close figure windows.');
  disp(' ');
```

Figure 6.9a. Main program to calculate temperatures in a small tunnel **(CUC30.m)**.

```
%  Subprogram to be used with CUC30.m                    soil30.m
%  Also requires function(s) FRSNL.m, FWS.m, TABS.m
%
%  Try various pp  values in'FRSNL.m' to see the effects of radiation.
%  Try various GSI values in'soil30.m' to see the effects of wetness.
%  Try various QH  values in 'soil30.m' to see the effects of airflow rate.
%
function dy = soil30(t,y)
global FL pp GSI QH
Tavg=15.0; TU=5.0; TBL=20.0; TD=8; % Temp (C)
%TD: outside dew point temperature, in degree C
KS = 5.5; CS= 2.0E+3; CA=1.164; CC=50.0;
% KS (kJ/m/C/hr) and KS/3.6 (W/m/C) also CS (kJ/m^3/C)
% CA: Volumetric heat capacity of air (kJ/m^3/C)
% CC: Heat capacity of cover (kJ/m^3/C)
RHO=1.164;GSI=1.0;SIG = 20.4;
% RHO: Density of air (kg/m^3)
% GSI: Greenhouse Soil's Wetness Index (1.0 is totally wet)
% SIG:Stefan-Boltzmann constant (kJ/m^2/K^4/hr) = 5.67(W/m^2/K^4)
HI=7.2; HO= 25.2;AH=0.5; QH=300;
% HI: Heat transfer coeff at soil surface (kJ/m^2/hr/C)
% HO: Heat transfer coeff at cover surface
% AH: Average air space height (m)
% QH: Air flow rate (m^3/m^2/hr)
Z0=0.01; Z1=0.05; Z2=0.1; Z3=0.2; Z4=0.5; %Z0-Z4: Depths of soil layers
(m)
ALC=0.1;ALF=0.7; RMC=0.1;RMSC=0.05;
EPSC=0.15; EPSF=0.95;
EPSA = 0.711+(TD/100)*(0.56+0.73*(TD/100));
TDS=1-ALC-RMSC;   TLV=1-EPSC-RMC;
HLG=2501.0; LE=0.9; KM=3.6*HI/LE/CA*RHO;
```

```
FL=1;           %Thickness of cover FL in mm
DEX=0.001*FL;   %Thickness of cover DEX in m
%
WO=FWS(TD);                              % calling function FWS()
clk = mod(t,24);
OMEGA=2*pi/24;                           % Time (hr)
TO= Tavg+TU*sin(OMEGA*(clk-8));
[TRAN,ABSO,RAD,RADS]=FRSNL(clk);         % calling function FRSNL()
%------------------------------------------------------------------
TC=y(1);TI=y(2);WI=y(3);TF=y(4);
T1=y(5);T2=y(6);T3=y(7);T4=y(8);
%------------------------------------------------------------------
% cover
TO4=TABS(TO);   TC4=TABS(TC);   TF4=TABS(TF);   % calling function TABS()
% Beginning of HWT calculation
HWT=0;
WSS=FWS(TC);                             % Calling function FWS()
if WSS < WI
   TW=WI-WSS;WI=WSS;
   HWT=TW*HLG*KM;            % Energy required to condense TW amount of water
End
% End of HWT calculation
ITC=(RAD*ABSO+RADS*ALC+(1-ALF)*(RADS*TDS+RAD*TRAN)*ALC...
   +SIG*EPSC*(EPSF*TF4 + EPSA*TO4+((1-EPSC)-2)*TC4)...
   -HO*(TC-TO)-HI*(TC-TI)+HWT)/DEX/CC;
%------------------------------------------------------------------
% Inside Air Layer
ITI=(HI*(TC-TI)+CA*QH*(TO-TI)+HI*(TF-TI))/CA/AH;
% ITI: net energy gain
% TI: Inside air temperature (C)
% WI: Inside humidity ratio (kg/kg)
WSTF=FWS(TF);
IWI=(RHO*QH*(WO-WI)+KM*(GSI*WSTF-WI))/RHO/AH;
% IWI: net mass gain
%------------------------------------------------------------------
%  At soil surface
ITF=(ALF*(RADS*TDS+RAD*TRAN)-KS*(TF-T1)*2/(Z0+Z1) ...
   -HI*(TF-TI)+HLG*KM*(WI-GSI*WSTF)...
   -SIG*EPSF*((1-RMC*EPSF)*TF4-EPSA*TLV*TO4-EPSC*TC4))/CS/Z0;
%------------------------------------------------------------------
% In soil layer
IT1 = (KS*(TF - T1)*2.0/(Z0+Z1)+KS*(T2 - T1)*2.0/(Z1+Z2))/CS/Z1;
IT2 = (KS*(T1 - T2)*2.0/(Z1+Z2)+KS*(T3 - T2)*2.0/(Z2+Z3))/CS/Z2;
IT3 = (KS*(T2 - T3)*2.0/(Z2+Z3)+KS*(T4 - T3)*2.0/(Z3+Z4))/CS/Z3;
IT4 = (KS*(T3 - T4)*2.0/(Z3+Z4)+KS*(TBL - T4)*2.0/Z4)/CS/Z4;
dy=[ITC; ITI; IWI; ITF; IT1; IT2; IT3; IT4]; % a column matrix
return
```

Figure 6.9b. Subprogram to calculate temperatures in a small tunnel (**soil30.m**).

If we change some constants that represent film properties, we can discuss the difference between covering materials such as PVC and PE using this model. Furthermore, this model can be used to simulate the environment in a greenhouse with a single layer cover if the average height (**AH**) of the air gap is increased. The first dynamic model of a greenhouse was reported by the author (Takakura et al., 1971), and its scheme is shown in Fig. 6.11. Two programs of this model, one in **FORTRAN** and the other in **MIMIC**, have been compared and are discussed from the viewpoint of simulation techniques (Takakura and Jordan, 1970).   This model was later translated into **CSMP** and compared with other models (van Bavel et al., 1987).   The main difference between the original model and the present model **CUC30** is that the heat flow in the soil was two-dimensional in the original model, as shown in Fig. 6.11.   Simulation study is quite advanced for the analysis of the greenhouse system; Takakura (1988 and 1989) gives literature reviews on this topic.

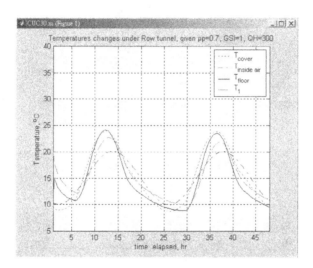

Figure 6.10a. Temperatures changes in a row tunnel

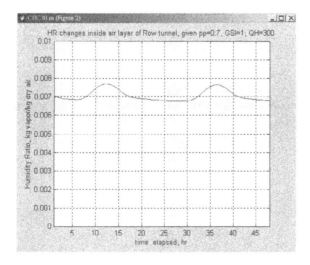

Figure 6.10b. Humidity ratio changes in a row tunnel.

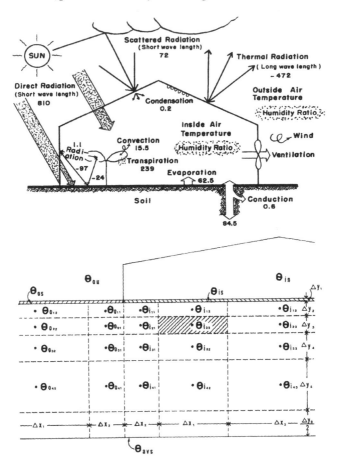

Figure 6.11. Heat flow components considered and their typical values in the dynamic model developed by Takakura et al. (1971).

6.4. DOUBLE LAYER GREENHOUSE MODEL (**CUC50**)

In this section we will discuss a greenhouse with two layers of cover.   Radiation exchanges around two layers of cover are rather complicated because three phases -- transmissivity, absorptivity, and reflectivity -- are involved.

The radiation exchange is shown in the diagram in Fig. 6.12.   From left to right in the figure, thermal radiation balance at the outer cover, at the inner cover, at the soil surface, and two types of short-wave radiation incoming to the soil surface, direct (**RAD**) and diffuse (**RADS**) radiation, are indicated.   The coefficients **TLV** and **TLVS** are the thermal radiation transmissivity of the outer cover and the inner screen, respectively.   **TDS** and **TDSS** are the absorptivity of the outer cover and the inner cover, respectively, for diffuse solar radiation, and the corresponding terms are **TRAN** and **TRANS** for direct solar radiation.   Radiation rays are indicated using arrows.   First order reflectance at the cover and screen is taken into consideration and shown by dotted arrows.   The temperature of the outer cover (**TC**), inner cover (**TS**) and floor (**TF**) are also indicated.

Combining these elements in radiation exchange form makes it possible to understand the energy balance equations in Fig. 6.13, which gives the whole program for an unheated greenhouse with a double layer cover.   For example, let us consider long wave radiation exchange at the outside cover, which is the left-most part of Fig. 6.12.   There are four incoming components: one is from the sky (**SIG * EPSA * TO4**), and three are from below the cover and directed upward -- Long wave radiation from the soil surface transmitted through the inside screen (**SIG * EPSF * TF4 * TLVS**), Long wave radiation emitted from the inside screen (**SIG * EPSS * TS4**), and radiation from the outside cover which is reflected back from the inside screen (**RMS * SIG * EPSC * TC4**).

There are two outgoing radiation components from the outside cover itself: one component is directed toward the sky, and the other component is directed downward (**2.0 * SIG * EPSC * TC4**).   These relationships are programmed in Fig. 6.13.   A close look at the **% Outer cover** part of the program shows that the incoming and outgoing long wave radiation components are combined, and this is expressed as **SIG * EPSC * (EPSF * TF4 * TLVS + EPSA * TO4 + EPSS * TS4 + (RMS * EPSC - 2.0) * TC4)** in the equation for **ITC** in the **% Outer cover** part.   All energy gains are shown with plus signs and losses with minus signs, and the absorptivity of the cover (**EPSC**) is multiplied to get net energy gain in the cover.   You will be able to understand the rest of the radiation balance in the same way.   Two lists of symbols, attached at the end of this chapter, were compiled to help the readers in understanding the program.   First is the list of symbols for all optical properties and second is a list of all symbols with descriptions.

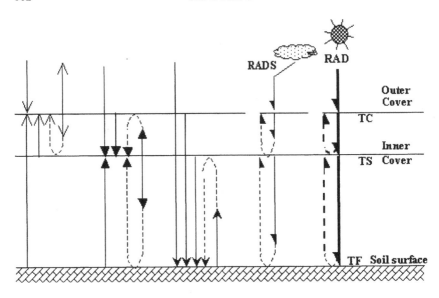

RADS   RAD

Outer
Cover

TC

Inner
TS   Cover

TF   Soil surface

Figure 6.12. Thermal radiation exchange around a double layer of covering (radiation balance of leaves is omitted).

```
% Double layer greenhouse model                              CUC50.m
% function(s): soil50.m
function cuc50(action)
if nargin==0, action='init';end
switch action
case 'init'
clc
global FL pp ARTO QH
tic
tini=0;tfinal=48;
y0=[10; 10; 0.01; 10; 15; 20; 20; 20; 20; 20]; % a column matrix
%   TC, TS,   WI,  TI,  TP, TF, T1, T2, T3, T4
[t,y]=ode23t('soil50',[tini tfinal],y0);
toc
%--------------------------------------------------------------------
h1=findobj('tag','cuc50_part1');close(h1);
figName='CUC50.m (Figure 1: Temperature changes ';
figName=[figName 'in a double-layer plastic house)'];
figure('tag','cuc50_part1','Resize','on','MenuBar','none',...
   'Name',figName,'NumberTitle','off','Position',[120,120,520,420]);
plot (t,y(:,1),'x-',t,y(:,2),'*-',t,y(:,4),'o-',t,y(:,5),'^-', ...
   t,y(:,6),'v-');
grid on;   axis([-inf, inf, 5, inf]);
titcont=['Given pp=' num2str(pp,3) ', ARTO=' num2str(ARTO,4) ...
   ', QH=' num2str(QH,4)];
title(titcont);xlabel('time elapsed, hr');  ylabel('Temperature, ^oC');
legend('T_C','T_S','T_I','T_P','T_F',-1);
%--------------------------------------------------------------------
h1=findobj('tag','cuc50_part2');close(h1);
figName='CUC50.m (Figure 2: (Under)ground soil temperature ';
figName=[figName 'changes in a double-layer plastic house]'];
```

```
figure('tag','cuc50_part2','Resize','on','MenuBar','none', ...
  'Name',figName,'NumberTitle','off','Position',[160,80,520,420]);
plot (t,y(:,6),'x-',t,y(:,7),'*-',t,y(:,8),'o-',t,y(:,9),'^-',t, ...
  y(:,10),'v-');
grid on; axis([-inf, inf, 5, inf]);
titcont=['Given pp=' num2str(pp,3) ', ARTO=' num2str(ARTO,4)...
    ', QH=' num2str(QH,4)];
title(titcont);
xlabel('time elapsed, hr'); ylabel('Temperature, ^oC');
legend('T_F','T1','T2','T3','T4',3);
%------------------------------------------------------------------
h1=findobj('tag','cuc50_part3');close(h1);
figName='CUC50.m (Figure 3: HR and Temperature ';
figName=[figName ' of air in a double-layer plastic house)'];
figure('tag','cuc50_part3','Resize','on','MenuBar','none','Name',...
  figName,'NumberTitle','off','Position',[200,40,520,420]);
subplot(2,1,1);    plot(t,y(:,3)); ylabel('Humidity Ratio, kg/kg DA');
grid on;
titcont=['Given pp=' num2str(pp,3) ', ARTO=' num2str(ARTO,4)...
    ', QH=' num2str(QH,4)];
title(titcont);
subplot(2,1,2); plot(t,y(:,4));
ylabel('Air Temperature, ^oC');   xlabel('time elapsed, hr');
grid on;
%
disp('Thank you for using CUC50.'); disp(' ');
disp('You can enter ''close all'...
    ' in the command window to close figure windows.');
disp(' ');
end % switch
```

Figure 6.13a. Main program to calculate temperatures in a plastic house with double layer covering (**CUC50.m**).

```
% Subprogram to be used with CUC50.m                      soil50.m
% Also require other functions: FRSNL.m, FWS.m, TABS.m
% by varying pp value in FRSNL.m,
%     the effect of radiation on T can be revealed.
%
function dy = soil50(t,y)
%
global FL pp QH ARTO
Tavg=15.0; TU=5.0; TBL=20.0; % in degree C
TD=8;               % outside dew point temperature, in degree C
KS = 5.5;           % KS is in kJ/m/C/hr and KS/3.6 is in W/m/C
CS= 2.0E+3;         % CS is in kJ/m^3/C
CA=1.164;           % Volumetric heat capacity of air (kJ/m^3/C)
CC=50.0;            % Heat capacity of cover (kJ/m^3/C)
RHO=1.164;          % Density of air (kg/m^3)
GSI=1;              % Greenhouse Soil's Wetness Index (1.0 is totally wet)
SIG = 20.4;         % Stefan-Boltzmann cst.(kJ/m^2/K^4/hr)=5.67W/m^2/K^4
QH=10.8;            % Air flow rate (m^3/m^2/hr), 8 cfm/ft^2 = 146 m^3/m^2/hr
                    % Case 1: QH=146; case 2: QH=10.8;
% Convective heat transfer coefficients (h): HO, HS and HI
HO= 25.2;           % h for outer cover facing upward
HS=10.8;            % h for outer cover facing downward
                    %   for inner cover facing upward.
                    %   for TB(inside covers' air T) facing up and down.
HI=7.2;             % h for inner cover facing downward (kJ/m^2/hr/C)
                    %   for plant surface (and floor) to inside air
```

```
                        %     for TI(air T in GH) facing up, down & facing plant
Z0=0.01; Z1=0.05; Z2=0.1; Z3=0.2; Z4=0.5;    % Depths of soil layer (m)
ALC=0.1;              % Absorptivity of cover for diffused solar radiation
ALF=0.7;              % Absorptivity of solar radiation at soil surface
ALS=0.1;              % Absorptivity of screen for diffused solar radiation
ALP=0.8;              % Absorptivity of plant for solar radiation
RMSC=0.05;            % Reflectivity, outer cover, diffused
RMSS=0.005;           % Reflectivity, inner cover, diffused
RMS=0.1;              % Reflectivity, outer cover, long wave
RMC=0.1;              % Reflectivity, inner cover, long wave
EPSA = 0.711+(TD/100)*(0.56+0.73*(TD/100));
                     % Emissivity, outside air,    long wave
EPSC=0.15;           % Emissivity, outer cover,    long wave
EPSS=0.15;           % Emissivity, inner cover,    long wave
EPSP=0.95;           % Emissivity, plant surface,  long wave
EPSF=0.95;           % Emissivity, soil surface,   long wave
TDS=1-ALC-RMSC;      % Transmissivity, outer cover, diffused
TDSS=1-ALS-RMSS;     % Transmissivity, inner cover, diffused
TLV=1-EPSC-RMC;      % Transmissivity, outer cover, long wave
TLVS=1-EPSS-RMS;     % Transmissivity, inner cover, long wave
HLG=2501.0; LE=0.9;
KM=3.6*HI/LE/CA*RHO;
AF=1000;             % AF:  Floor area, m^2
AH=2;                % AH:  Average air space height (house height, m)
AP=1;                % AP:  Projected plant area (m^2) per plant
NP=1;                % NP:  number of plants (NP*AP should not > AF)
                     % Case 1: NP=600; case 2: NP=1;
VP=1;                % VP:  Volume of crop per plant
LAI=1;               % LAI: Leaf area index per plant
CP=4180;             % CP:  heat capacity of plant, (kJ/m3/C)
GAM=0.132;           % GAM: water vapor diffusion resistance (hr/m)
FL=1;                % Thickness of cover in mm
DEX=0.001*FL;        % Thickness of cover in m
WofTD=FWS(TD);       % calling function FWS()
ARTO=NP*AP/AF;       % Ratio of total plant area to floor area
ARTO1=1-ARTO;        % Ratio of non-plant area to floor area
%-------------------------------------------------------------------
clk = mod(t,24); OMEGA=2*pi/24;     TO= Tavg+TU*sin(OMEGA*(clk-8));
[TRAN,ABSO,RAD,RADS]=FRSNL(clk);          % calling function FRSNL()
%-------------------------------------------------------------------
% Outer Cover
TC=y(1);TS=y(2);WI=y(3);TI=y(4);TP=y(5);
TF=y(6);T1=y(7);T2=y(8);T3=y(9);T4=y(10);
TO4=TABS(TO);        % calling function TABS()
TC4=TABS(TC);        % TC is T of outside cover
TS4=TABS(TS);        % TS is T of inside screen
TP4=TABS(TP);        % TP is T of plant surface
TF4=TABS(TF);        % TF is T of floor (soil surface)
TB=(TC+TS)/2;        % TB is average T of TC and TS
TRANS=TRAN;  ABSOS=ABSO;
ITC=(RAD*ABSO+RADS*ALC+((1-TRANS-ABSOS)*RAD...
  *TRAN+RMSS*RADS*TDS)*ALC+SIG*EPSC*(EPSF...
  *TF4*TLVS+EPSA*TO4+EPSS*TS4+(RMS-2)*TC4)...
  +HO*(TO-TC)+HS*(TB-TC))/DEX/CC;
%-------------------------------------------------------------------
% Inner Cover
HWT=0;
ITS=(RAD*ABSOS*TRAN+RADS*TDS*ALS+(1-ALF)*(RADS...
  *TDS*TDSS+RAD*TRAN*TRANS)*ALS+HS*(TB-TS)+HI...
  *(TI-TS)+SIG*EPSS*(EPSF*TF4+EPSA*TLV*TO4...
  +EPSC*TC4-(2-RMC-(1-EPSF))*TS4)+HWT)/DEX/CC;
```

```
%
WSS=FWS(TC);              % Calling function FWS()
if WSS < WI
   TW=WI-WSS;             % amount of water condensed
   WI=WSS;
   HWT=TW*HLG*KM;         % Energy required to condense TW amount of water
end
%------------------------------------------------------------------------
% Inside Air Layer
ITI=(HI*(TS-TI)+CA*QH*(TO-TI)+HI*(TP-TI)*ARTO*LAI...
   *2+HI*(TF-TI))/CA/AH;
%------------------------------------------------------------------------
WSTP=FWS(TP);   WSTF=FWS(TF);
IWI=(RHO*QH*(WofTD-WI)+KM*(GSI*WSTF-WI)+2*LAI*...
   ARTO*RHO/GAM*(WSTP-WI))/RHO/AH;
%------------------------------------------------------------------------
%   At plant surface
ITP=(2*LAI*HI*(TI-TP)+ALP*(RADS*TDS*TDSS+RAD*...
   TRAN*TRANS)+SIG*EPSP*(EPSA*TLV*TLVS*TO4+...
   EPSS*TS4-(1-RMS)*TP4+EPSC*TLVS*TC4)...
   -RHO*HLG/GAM*(WSTP-WI)*2*LAI)/CP/VP*ARTO;
%------------------------------------------------------------------------
%   At soil surface
ITF=(ALF*(RADS*TDS*TDSS+RAD*TRAN*TRANS)*ARTO1...
   -KS*(TF-T1)*2/(Z0+Z1)  ...
   -HI*(TF-TI)+HLG*KM*(WI-GSI*WSTF)...
   -SIG*EPSF*((1-RMS)*TF4-EPSA*TLV*TLVS*TO4...
   -EPSS*TS4-EPSC*TLVS*TC4))/CS/Z0;
%------------------------------------------------------------------------
% In Soil layer
IT1 = (KS*(TF - T1)*2/(Z0+Z1)+ KS*(T2 - T1) *2/(Z1+Z2))/CS/Z1;
IT2 = (KS*(T1 - T2)*2/(Z1+Z2)+ KS*(T3 - T2) *2/(Z2+Z3))/CS/Z2;
IT3 = (KS*(T2- T3) *2/(Z2+Z3)+ KS*(T4 - T3) *2/(Z3+Z4))/CS/Z3;
IT4 = (KS*(T3 - T4)*2/(Z3+Z4)+ KS*(TBL - T4)*2/Z4)/CS/Z4;
dy=[ITC; ITS; IWI; ITI;ITP; ITF; IT1; IT2; IT3; IT4];
return
```

Figure 6.13b. Subprogram to calculate temperatures in a plastic house with double layer covering (**soil50.m**).

Two separate simulation results are given in Figs. 6.14a, b and c. The left figures of Figs. 6.14a, b and c are results using **ARTO**=0.6 and **QH**=146 m³/m²/hr, representing 60% coverage of plants in the greenhouse and moderate ventilation, and the right figures are results using **ARTO**=0.001 and **QH**=10.8 m³/m²/hr, representing almost no plants in the greenhouse and limited air exchange. **ARTO** represents the degree of coverage of the projected area of plants in the greenhouse and is calculated by **NP * AP / AF**. **QH** is the airflow rate in metric units.

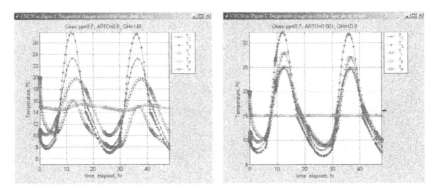

Figure 6.14a. Temperature changes of outer and inner cover, inside air, plant surface and floor in a plastic house with double layer s at various ARTO and QH when PP equals 0.7.

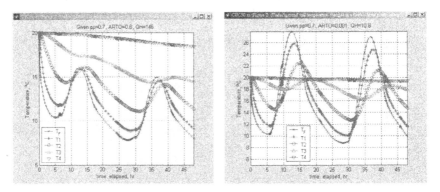

Figure 6.14b. Temperature changes of floor and soil layers in a plastic house with double layers at various ARTO and QH when PP equals 0.7.

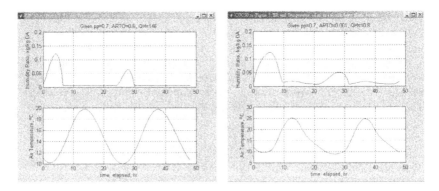

Figure 6.14c. Humidity ratio and temperature changes of air in a plastic house with double layers at various ARTO and QH when PP equals 0.7.

Fig. 6.14a shows temperature changes of the outer cover, inner cover, inside air, plant surface and soil surface.   Fig. 6.14b shows temperature changes the of floor and soil layers, and Fig. 6.14c shows Humidity ratio and temperature changes of air in a plastic house with double layers.   For ease of comparison, some data were shown more than once among the three figures.

As shown in Fig. 6.14a, Temperatures of the inner cover have greatest variation throughout the day.   The maximum and minimum temperatures of the cover are due to solar radiation in the daytime and radiative cooling at night.   It might be strange to have inside air temperature lower than outside; the phenomenon may be a bit exaggerated in the simulation results, but it is what often happens on cold nights. The main heat source for unheated greenhouses is solar radiation absorbed into the soil layer during the daytime.   A double layer cover reduces incoming daytime energy at the soil surface more than does a single layer.   The better insulation that a double layer provides at night cannot compensate.   However, higher wind speed, higher dew point air temperature, and more cloudiness will increase the inside air temperature.   Wind speed changes the heat transfer coefficient at the outside surface as a result of convection.   Long wave radiation from the sky varies with dew point temperature and cloudiness, and these factors can easily be incorporated into the model.

6.5. A PAD AND FAN GREENHOUSE MODEL (CUC35)

There are several types of evaporative cooling systems being used in practical greenhouses.   The pad and fan system is one type of evaporative cooling system that is very common in area with hot and dry weather conditions.   On the psychrometric chart (in Fig. 4.8), the process can be shown by following along the contour line of wet bulb temperature of the air.   When water is sprayed into the air, it is vaporized and the amount of water used is not large.   The whole process is considered as adiabatic and the process ends in an ideal condition at the saturation point (relative humidity is 100%), which is when the dry-bulb temperature is equal to the wet-bulb temperature.   Normally 70-80% of the total process can be reached.

In a pad and fan greenhouse, for example, a pad is installed on one end of a greenhouse wall.   Water is sprayed from the top of the pad and the air is sucked in through the pad by fans that are installed at the other end of the greenhouse wall. The air flows horizontally from one side of the greenhouse to the other.   Air temperature and humidity gradients exist along the greenhouse.   One pair of air temperature and humidity measurements cannot represent the inside condition well. A model simulate this type of situation can be built by modifying the model of a single greenhouse or a row tunnel (CUC30).   The model in program CUC35 includes two-dimensional heat flow.   The basic parts are shown in Fig. 6.15.

The greenhouse is divided into three equal parts.   Normally air is sucked in through the pad, flows through the greenhouse, and is exhausted through the fan. Air temperature is expressed by TI1, TI2 and TI3, and humidity is expressed by humidity ratios WI1, WI2, and WI3.   In this example, the temperature of the glass covering (TC) is uniform, and is the same as the surface temperature of the soil

(TF). Temperature and humidity conditions immediately after the air passes through the pad are **TOI** and **WOI**, respectively. These are determined by the psychrometric properties of the air and the efficiency of the pad and fan system.

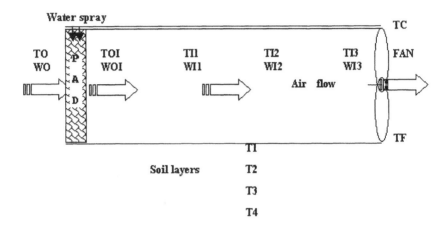

Figure 6.15. Diagram of a pad and fan greenhouse

The model solves 12 unknown variables simultaneously as shown in eq. 6.2 and listed in Fig. 6.16b. The initial conditions of these variables are shown in eq. 6.3 and listed in Fig. 6.16a. The dimension of **dy** and **y0** arrays/matrices must be identical. As shown in eqs. 6.2 and 6.3, the ' sign after the right bracket of matrices **dy** and **y0** is the command to take the transpose of a matrix.

$$\mathbf{dy} = [\text{ITC ITF ITI1 ITI2 ITI3 IWI1 IWI2 IWI3 IT1 IT2 IT3 IT4}]' \quad (6.2)$$

$$\mathbf{y0} = [10\ 10\ 10\ 10\ 10\ 0.01\ 0.01\ 0.01\ 20\ 20\ 20\ 20]' \quad (6.3)$$

To solve **TC**, **TF,** and **T1** to **T4**, there are no differences between this model and **CUC30**. However, for the temperatures and humidity ratios within the greenhouse, some modifications are required. First, define **TI1, TI2, TI3, WI1, WI2,** and **WI3** using eq. 6.4 as an example (see corresponding equation in Fig. 6.16B). Eq. 6.4 is used to solve for **TI1**, which is the air temperature of the first 1/3 part of the greenhouse. Note that the integral of **ITI1** is in fact **TI1**.

$$\text{ITI1} = \ (\text{HI} * (\text{TC - TI1}) + \text{CA} * \text{QH} * ((\text{TOI - TI1}) -$$
$$(\text{TI1} - \text{TI2})) + \text{HI} * (\text{TF - TI1})) / \text{CA} / \text{AH} \quad (6.4)$$

where **QH** is the air flow rate $(\text{m}^3/\text{hr/m}^2)$ on a unit floor area basis and can be a function of time.

Second, change the boundary conditions **TC** and **TF**.   To generate the equation for the **ITC** term, the **–HI * (TC - TI)** term is replaced by **-HI * (TC –TIaverage)**, where **TIaverage** equals **(TI1 + TI2 + TI3)** / 3. Also **+HWT** is replaced by **(+ HWT1 + HWT2 + HWT3)**.

Third, change the condensation term.   For ·each of the three regions, comparisons should be made between the humidity ratio of the covering temperature (**TC**) and that of the air temperature (**TI**).   Create a separate function **conds** (listed in Fig. 6.16b) to facilitate comparison of humidity ratios.

Fourth, remove the **HWT** calculation section (from **% Beginning of HWT calculation** to **% End of HWT calculation** in the 'soil30.m' program listed in Fig. 6.9b) and include the following:

HWT1 = CONDS(TC, WI1)

HWT2 = CONDS(TC, WI2)

HWT3 = CONDS(TC, WI3)

In this model, outside dew point temperature is one of the inputs.   The physical properties of outside air then can be determined by dry bulb and dew point temperatures.   One of the reasons for using dew point is that it is fairly constant throughout the day, and during initial testing, we can assume it is constant. Secondly, in this model the emissivity of the air is expressed as a function of dew point temperature.   When at least two properties of the air are given, the remaining properties can be determined.   In this case, first, humidity ratio (**WO**) is calculated.

Using the **wbcal** function, a regression equation solving wet-bulb temperature (**TOW**) as functions of dry-bulb and dew point temperature, the wet-bulb temperature is calculated.   Dry-bulb temperature is always greater than or equal to wet bulb temperature, which is again always greater than or equal to dew point temperature.   The three temperatures are the same when the relative humidity equals 100%.

Intermediate inputs **TOI** and **WOI** are given as the functions of **TO** and **WO** using the psychrometric relationships and the efficiency of the pad (**PEFF**).   **TOI** and **WOI** can be obtained using the following relations:

TOI   = TO – PEFF * (TO - TOW)

WOI = WO + PEFF * (WWS - WO)

In evaporative cooling systems such as pad and fan systems, incoming air is cooled along the wet-bulb contour line on the psychrometric chart.   If the efficiency is 100% (1 in decimal notation), **TOI** equals **TOW** and **WOI** equals **WWS**, which means the incoming air temperature drops to the wet bulb temperature and the humidity is 100%.   This is the maximum efficiency situation.

```
% Pad-and-fan greenhouse model                                    CUC35.m
% Also requires function: soil35.m
%
function cuc35(action)
if nargin==0, action='init';end
clc
fprintf('\n\nPress Enter to continue\n\n');
h1=findobj('tag','wait');close(h1);
figure('tag','wait','Resize','off','MenuBar','none','Name', ...
   'Please wait.','NumberTitle','off','Position',[300,300,160,80], ...
   'color',[0.8 0.8 0.8]);
h2a=uicontrol('style','text','string', ...
   'This program will take a while to run.  Please switch to '...
   'Command Window'', then press <Enter> to start.',...
   'position',[10,5,140,70],'backgroundcolor',[0.8 0.8 0.8]);
pause
%
switch action
case 'init'
clc
%
t0=0;tfinal=48;
y0=[10 10 10 10 10 0.01 0.01 0.01 20 20 20 20]';
%   TC  TF  TI1 TI2 TI3 WI1   WI2   WI3  T1 T2 T3 T4
%   List of initial conditions
%
[t,y]=ode15s('soil35',[t0 tfinal],y0);
%
h1=findobj('tag','wait');close(h1);
h1=findobj('tag','cuc35_part1');close(h1);
figure('tag','cuc35_part1','Resize','on','MenuBar','none', 'Name',...
   'CUC35.m (Fig1: Temp. changes in a single layer pad & fan plastic house)',...
   'NumberTitle','off','Position',[120,120,520,420]);
%-[Fig.1]----------------------------------------------------------
% regenerate TO data for Figure 1
t1=0:1:48;
OMEGA=2*pi/24;  % Time (hr)
Toutdoor= 30+5*sin(OMEGA.*(mod(t1,24)-8));
plot (t1,Toutdoor,'r-',t,y(:,3),'b-.',t,y(:,4),'k--',t,y(:,5),'m:');
grid on;
axis([-inf, inf, 10, 40]);
xlabel('time elapsed, hr');
ylabel('Temperature, ^oC');
legend('TO','TI1', 'TI2', 'TI3',4);
%-[Fig.2]----------------------------------------------------------
h1=findobj('tag','cuc35_part2');close(h1);
figure('tag','cuc35_part2','Resize','on','MenuBar','none', 'Name',...
   'CUC35.m (Fig2: Soil T changes in a single layer pad & fan plastic house)',...
   'NumberTitle','off','Position',[160,80,520,420]);
plot (t,y(:,1),t,y(:,2),t,y(:,9),t,y(:,10),t,y(:,11),t,y(:,12));
grid on;
axis([-inf, inf, 5, inf]);
xlabel('time elapsed, hr');
ylabel('Temperature, ^oC');
legend('T_c_o_v_e_r','T_f_l_o_o_r','T_1','T_2','T_3','T_4',4);
%-[Fig.3]----------------------------------------------------------
h1=findobj('tag','cuc35_part3');close(h1);
figure('tag','cuc35_part3','Resize','on','MenuBar','none', 'Name',...
   'CUC35.m (Fig3: Humidity Ratio in a single layer pad & fan plastic
```

```
house)',...
  'NumberTitle','off','Position',[200,40,520,420]);
plot (t,y(:,6),'r-',t,y(:,7),'b:',t,y(:,8),'k-.');
grid on;
legend('WI1','WI2','WI3',4);
xlabel('time elapsed, hr');
ylabel('Humidity Ratio, kg vapor/kg dry air');
%
fprintf('\n\n');
%
  disp('Thank you for using CUC35');
  disp('You can enter ''close all''…
     ' in the command window to close figure windows.');
  disp(' ');
end      % switch
```

Figure 6.16a. Main program of the pad and fan greenhouse model **(CUC35.m)**

```
% Subprogram to be used with CUC35.m                     soil35.m
% Also requires functions FRSNL.m, FWS.m, TABS.m,
%             and functions conds() and wbcal() (in this file)
% Major inputs:
%      location (latitute and longitude)
%      Time of the year from Table 6.1
%      Extinction coefficient (pp) for solar radiation
%      outside dry-bulb temperature (TO) and dew point temperature (TD)
function dy = soil35(t,y)
global HLG KM FL pp ARTO
%
Tavg=30.0; TU=5.0; TBL=20.0; TD=15; % Temp (C)
%TD: outside dew point temperature, in degree C
KS = 5.5; CS= 2.0E+3; CA=1.164; CC=50.0;
% KS (kJ/m/C/hr) and KS/3.6 (W/m/C) also CS (kJ/m^3/C)
% CA: Volumetric heat capacity of air (kJ/m^3/C)
% CC: Heat capacity of cover (kJ/m^3/C)
RHO=1.164; % Density of air (kg/m^3)
GSI=1;       % Greenhouse Soil's Wetness Index (1.0 is completely saturated)
SIG = 20.4;
% SIG:Stefan-Boltzmann constant (kJ/m^2/K^4/hr) = 5.67(W/m^2/K^4)
QH=36;       % Ventilation air flow rate (m^3/m^2/hr), 8 cfm/ft^2 = 146 m/hr
AH=2;        % Average air space height (house height, m)
HO= 25.2;    % h at cover surface facing upward
HI=7.2;      % h at cover surface facing downward (kJ/m^2/hr/C)
Z0=0.01; Z1=0.05; Z2=0.1; Z3=0.2; Z4=0.5; % Depths of soil layers (m)
ALC=0.1;     % Absorptivity of cover for diffused solar radiation (ND)
ALF=0.7;     % Absorptivity of solar radiation at soil surface
RMC=0.1;     % Reflectivity of screen for long wave radiation
RMSC=0.05;   % Reflectivity of cover for diffuse solar radiation
EPSC=0.15;   % Emissivity of cover
EPSF=0.95;   % Emissivity of soil surface
HLG=2501.0;LE=0.9; KM=3.6*HI/LE/CA*RHO;
FL=3;   DEX=0.001*FL;   %Thickness of cover FL in mm and DEX in m
PEFF=0.8;    % Efficiency of pad and fan system
EPSA = 0.711+(TD/100)*(0.56+0.73*(TD/100));
% EPSA:Emissivity of air layer
WO=FWS(TD);% WO: humidity ratio of outdoor air=saturated V.P. at Tdp
TDS=1-ALC-RMSC;
% TDS: Transmissivity of cover for diffuse solar radiation
TLV=1-EPSC-RMC;    % Transmissivity of cover for long wave radiation
%-------------------------------------------------------------------
```

```
clk = mod(t,24);   OMEGA=2*pi/24;    % Time (hr)
TO= Tavg+TU*sin(OMEGA*(clk-8));      % TO is outdoor air temperature
[TRAN,ABSO,RAD,RADS]=FRSNL(clk);     % calling function FRSNL()
%-------------------------------------------------------------------
% Pad and fan system
TOW=wbcal(TO,TD);       % calling function wbcal to calculate Twb
WWS=FWS(TOW);           % calling function FWS()
TOI=TO-PEFF*(TO-TOW);  WOI=WO+PEFF*(WWS-WO);
% TOI: Dry bulb T right after pad (in oC)
% WOI: Humidity ratio right after pad (in kg vapor/kg dry air)
%-------------------------------------------------------------------
TC=y(1); TF=y(2); TI1=y(3);TI2=y(4); TI3=y(5);
WI1=y(6);WI2=y(7);WI3=y(8); T1=y(9);T2=y(10); T3=y(11);T4=y(12);
%-------------------------------------------------------------------
% Cover
TO4=TABS(TO);  TC4=TABS(TC);  TF4=TABS(TF);  % calling function TABS()
WSS=FWS(TC);
% calculating HWT [HWT WII]
[HWT1, WI1]=conds(WSS,WI1);   [HWT2, WI2]=conds(WSS,WI2);
[HWT3, WI3]=conds(WSS,WI3);
avgTCTIdif=(TC-(TI1+TI2+TI3)/3);
sumHWT=HWT1+HWT2+HWT3;
ITC=(RAD*ABSO+RADS*ALC+(1-ALF)*(RADS*TDS...
   +RAD*TRAN)*ALC+SIG*EPSC*(EPSF*TF4...
   +EPSA*TO4+((1-EPSF)-2)*TC4) ...
   -HO*(TC-TO)-HI*avgTCTIdif+sumHWT)/DEX/CC;
fprintf('t=%4.2f TO=%4.2f TOI=%4.2f WOI=%4.2f sumHWT=%4.2f\n',t,TO, ...
   TOI,WOI,sumHWT);
%-------------------------------------------------------------------
% Inside Air Layer
WSTF=FWS(TF);
ITI1=(HI*(TC-TI1)+CA*QH*((TOI-TI1)-(TI1-TI2))+HI*(TF-TI1))/CA/AH;
IWI1=(RHO*QH*((WOI-WI1)-(WI1-WI2)+KM*(GSI*WSTF-WI1)))/RHO/AH;
ITI2=(HI*(TC-TI2)+CA*QH*((TI1-TI2)-(TI2-TI3))+HI*(TF-TI2))/CA/AH;
IWI2=(RHO*QH*((WI1-WI2)-(WI2-WI3)+KM*(GSI*WSTF-WI2)))/RHO/AH;
ITI3=(HI*(TC-TI3)+CA*QH*((TI2-TI3)-(TI3-TO))+HI*(TF-TI3))/CA/AH;
IWI3=(RHO*QH*((WI2-WI3)-(WI3-WO)+KM*(GSI*WSTF-WI3)))/RHO/AH;
%-------------------------------------------------------------------
%   At soil surface
avgTITFdif=(TF-(TI1+TI2+TI3)/3);
avgWI=((WI1+WI2+WI3)/3);
ITF=(ALF*(RADS*TDS+RAD*TRAN)-KS*(TF-T1)*2/(Z0+Z1) ...
   -HI*avgTITFdif+HLG*KM*(avgWI-GSI*WSTF)...
   -SIG*EPSF*((1-RMC)*TF4-EPSA*TLV*TO4...
   -EPSC*TC4))/CS/Z0;
%-------------------------------------------------------------------
% In Soil layer
IT1 = (KS*(TF-T1)*2/(Z0+Z1)+KS*(T2-T1)*2/(Z1+Z2))/CS/Z1;
IT2 = (KS*(T1-T2)*2/(Z1+Z2)+KS*(T3-T2)*2/(Z2+Z3))/CS/Z2;
IT3 = (KS*(T2-T3)*2/(Z2+Z3)+KS*(T4-T3)*2/(Z3+Z4))/CS/Z3;
IT4 = (KS*(T3-T4)*2/(Z3+Z4)+KS*(TBL-T4)*2/Z4)/CS/Z4;
dy=[ITC ITF ITI1 ITI2 ITI3 IWI1 IWI2 IWI3 IT1 IT2 IT3 IT4]';
return
%-------------------------------------------------------------------
function [HWT, WII]=conds(WWS,WII)
global HLG KM
HWT=0;
if WWS < WII
   TW=WII-WWS;       % amount of water condensed
   WII=WWS;
   HWT=TW*HLG*KM/3;% Energy required to condense TW amount of water
```

```
                    % 3 is the number of evenly divided compartments.
end
%-------------------------------------------------------------------
function twb=wbcal(tdb,tdp)
if abs(tdb - tdp) < .001, twb = tdb;break;end
 X = tdb;
 a = .011569; b = .613423862;  c = -.00643928; D = 7.52158e-05;
 e = -4.5287e-07;
 ap1 = a + b * X + c * X ^ 2 + D * X ^ 3 + e * X ^ 4;
 a = .419636669; b = .027436851; c = .007711576; D = .001536155;
 e = .00023861;
 bp1 = (a + c * X + e * X ^ 2) / (1 + b * X + D * X ^ 2);
 a = .011146403;b = .027956528; c = .000255119;D = .002122386;
 e = 7.1215e-06;
 cp1 = (a + c * X + e * X ^ 2) / (1 + b * X + D * X ^ 2);
 a = 9.65426e-05;b = -.00292091; c = 7.15163e-07;D = .001201577;
 dp1 = (a + c * X) / (1 + b * X + D * X ^ 2);
 twb = ap1 + bp1 * tdp * tdp + cp1 * tdp * tdp + dp1 * tdp * tdp * tdp;
if twb > tdb,twb = tdb;end
if twb < tdp,twb = tdp;end
```

Figure 6.16b. Subprogram of CUC35 model (**soil35.m**).

Fig. 6.17 shows results of the pad and fan (**CUC35**) model.   In total, 3 figures were generated.   Fig. 6.17a shows the outdoor and indoor air temperatures.   In the first 1/3 of the greenhouse, **TI1** is lower than the outdoor temperature, which due to the evaporative cooling effect of the pad.   Between 10 am to 3 pm, the air temperature at the last 1/3 (**TI3**) is higher than **TO**.   Try to rerun the model with 50% shading by lowering transmissivity of the cover, or by doubling the air speed inside the greenhouse by ventilation..

Fig. 6.17b shows the cover temperature, the floor temperature and the 4 layers of soil temperatures.   Due to the buffering effect of the air and the soil, the peak of each curves shifted to the right (time delay) can be clearly observed.   In the legend of this figure, the subscript letters can be generated using a leading '_' before each character as shown in the '%-[Fig.2]-' portion of Fig. 6.16a.   The curves in this figure in mono-color cannot be distinguwished well, but this problem could be solved by color output specification.

Fig. 6.17c shows the humidity ratio inside greenhouse.   Only little differences among **WI1, WI2 and WI3** can be observed during peak hours.   The increase of the humidity ratio inside the greenhouse is the result of the evapotranspiration of the plants. In the program listed in Fig. 6.16b, Greenhouse Soil's Wetness Index (**GSI**) equals 1 (saturation) is assumed.   Try to run the program with small value to see the differences among **WI1**, **WI2**, and **WI3**.

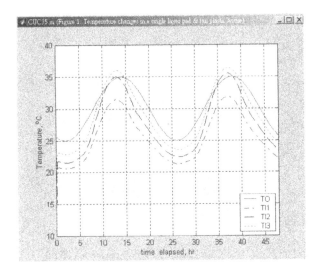

Figure 6.17a.The first output of the pad and fan model **(CUC35)**.

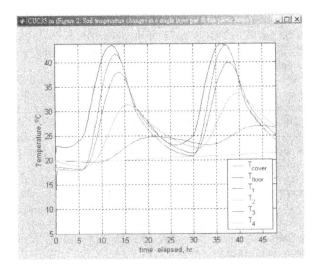

Figure 6.17b.The second output of the pad and fan model **(CUC35)**.

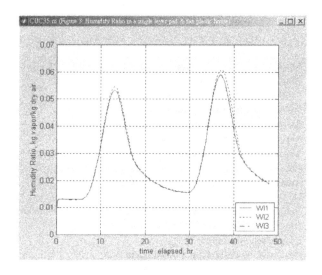

Figure 6.17c.The third output of the pad and fan model **(CUC35)**.

LIST OF SYMBOLS FOR OPTICAL PROPERTIES OF AIR, COVER, PLANT AND FLOOR FOR MODELS IN CHAPTER 6

	Absorptivity		*Emissivity*	*Reflectivity*		*Transmissivity*		
	direct	*diffuse*	*long wave*	*diffuse*	*long wave*	*direct*	*diffuse*	*long-wave*
Outside Air			**EPSA**					
Outer Cover	**ABSO**	**ALC**	**EPSC**	**RMSC**	**RMC**	**TRAN**	**TDS**	**TLV**
Inner Cover	**ABSOS**	**ALS**	**EPSS**	**RMSS**	**RMS**	**TRANS**	**TDSS**	**TLVS**
Plant	**ALP**	**ALP**	**EPSP**					
Floor	**ALF**	**ALF**	**EPSF**			**0**	**0**	**0**

LIST OF SYMBOLS WITH TYPICAL VALUES FOR MODELS IN CHAPTER 6

ABSO	Absorptivity of the outer cover for direct solar radiation, Calculated (ND)
AF	Floor area (Cover area), (m^2)
AH	Average height of a greenhouse, (m)
AP	Projected area of single plant, (m^2)
ALF	Absorptivity of the floor for solar radiation, 0.8 (ND)
ALP	Absorptivity of the plant for solar radiation, 0.8 (ND)
ALC	Absorptivity of the outer cover for diffuse solar radiation, 0.1 (ND)
ALS	Absorptivity of the inner cover for solar radiation, 0.1 (ND)
ARTO	Area ratio of total plant leaf area to floor area, 0 (ND)
ARTO1	1.0 – ARTO, 1.0 (ND)
CA	Volumetric heat capacity of air at constant pressure, 1.164 (kJ/m^3/°C)
CC	Heat capacity of covers, 50.0 (kJ/m^3/°C)
CLOCK	Hour of day, 0 - 24 (hr)
CS	Heat capacity of the floor, 2000.0 (kJ/m^3/°C)
CP	Heat capacity of single plant, 4180.0 (kJ/m^3/°C)
CWP	Heat capacity of water, 4180.0 (kJ/m^3/°C)
DEC	Declination angle of the place, Input variable (deg)
DEX	Thickness of covers, 0.0001 (m)
EPSA	Effective emissivity of air, empirical equation after Brunt, (ND)
EPSC	Emissivity of outer cover, 0.15 (ND)
EPSF	Emissivity of the floor, 0.95 (ND)
EPSP	Emissivity of the plant, 0.95 (ND)
EPSS	Emissivity of inner cover, 0.15 (ND)
FK	Extinction coefficient, 0.0441 (1/mm)
FL	Thickness of the cover, 1 (mm)
FN	Index of refraction, 1.526 (ND)
GAM	Water vapor resistance, 0.132 (hr/m)
GSI	Greenhouse soil wetness index; 0 - 1.0 (ND)
HAG	Hour angle, Calculated (deg)
HF	Convective heat transfer coefficient at the floor, = HI (W/m^2/°C)
HI	Convective heat transfer coefficient at the inner cover, 5.0 (W/m^2/°C)
HLG	Latent heat for evaporation, 2501.0 (kJ/kg)
HO	Convective heat transfer coefficient at the outer cover, (W/m^2/°C)
HS	Convective heat transfer coefficient between covers, (W/m^2/°C)
HWT	Latent heat due to condensation on the inner surface of the cover, Calculated (kJ/m^2/hr)
J0W	Solar constant, 1360 (W/m^2)
K	Thermal conductivity of the floor, (W/m/°C)
KM	Mass transfer coefficient, 3.6*HI/LE/CA*RHO (kg/m^2/hr)
KS	Soil thermal conductivity, 5.5 (kJ/m/°C/hr)
LAI	Leaf area index, 1.0 (ND)
LATD	Latitude of the place, Input variable (deg)

LE	Lewis number, 3.6*HI/(KM*CA/RHO), (ND)
NP	Number of plants per unit area, 1 (1/m^2)
OMEGA	Frequency of time, 2x3.14/24 (1/hr)
PP	Extinction coefficient of the atmosphere, Input variable (ND)
RAD	Direct solar radiation, Calculated (W/m^2)
RADS	Scattered solar radiation, Calculated (W/m^2)
RHO	Density of air, 1.164 (kg/m^3)
RMC	Reflectivity of the outer cover for long wave radiation, 0.1 (ND)
RMS	Reflectivity of the inner cover for long wave radiation, 0.1 (ND)
RMSC	Reflectivity of the outer cover for diffuse solar radiation, 0.05 (ND)
RMSS	Reflectivity of the inner cover for diffuse solar radiation, 0.05 (ND)
SALT	Sun's altitude, Calculated (rad)
SO	Outside hemispherical solar radiation, Input variable (W/m^2)
SIG	Stefan-Boltzmann Constant, 5.67x10^{-8} (W/m^2/K^4)
TB	Air temperature between covers, Output variable ($^{\circ}$C)
TBL	Lower boundary of soil temperature, Input variable ($^{\circ}$C)
TC	Temperature of outer cover, Output variable ($^{\circ}$C)
TD	Dew point temperature of outside air, Input variable ($^{\circ}$C)
TDS	Transmissivity of cover for diffuse solar radiation, Input variable (ND)
TDSS	Transmissivity of screen for diffuse solar radiation, Input variable (ND)
TF	Temperature at the floor surface, Output variable ($^{\circ}$C)
TI	Inside air temperature, Output variable ($^{\circ}$C)
TLV	Transmissivity of cover for long wave radiation, Input variable (ND)
TLVS	Transmissivity of screen for long wave radiation, Input variable (ND)
TO	Outside air temperature, Input variable = **T0 + TU * sin(OMEGA*clock)** ($^{\circ}$C)
TP	Plant temperature, Output variable ($^{\circ}$C)
TS	Temperature of inner cover, Output variable ($^{\circ}$C)
TRAN	Transmissivity of the covering material, calculated (ND)
TT	Temperature of feeding warm water, Input variable ($^{\circ}$C)
T1 - T4	Soil layer temperature, Calculated ($^{\circ}$C)
VG	Greenhouse volume, (m^3)
VP	Volume of single plant, (m^3)
WI	Humidity ratio in the greenhouse, Output variable (kg/kg dry air)
WO	Outside humidity ratio, Calculated from TD (kg/kg dry air)
Z0 - Z4	Depth of each soil layer, Input variable (m)

PROBLEMS

1. In Europe, a kind of glass with low emissivity is available. One side of the glass sheet is metal-coated and its emissivity is 0.25 instead of 0.95. Modify the program **CUC30** and find which side should be metal-coated.

2. From the simulation result shown in Fig. 6.14, explain why higher wind speed, higher dew point air temperature and more cloudiness will increase the inside air temperature.

3. Change the program **CUC20** for the condition that the input is the wet-bulb temperature instead of dew-point temperature.

4. Modify the **CUC35** model to have five equal regions of air space in the greenhouse along the direction of the airflow.

5. If the case 1 greenhouse is to be renovated to reduce the inside temperature in the summer, what would your selection for the new design among cases 2, 3 and 4. Case 2 represents shading 50%, case 3 represents doubling ventilation rate and case 4 represents doubling the greenhouse height. In the practical situation, the cost for renovation should be considered. Rerun **CUC35** model using the following values.

    Case 1: No shading,   QH=3.6,   AH=2
    Case 2: 50% shading,   QH=3.6,   AH=2
    Case 3: No shading,   QH=7.2,   AH=2
    Case 4: No shading,   QH=3.6,   AH=4

CHAPTER 7

CO_2 ENVIRONMENT

7.1. INTRODUCTION

Carbon dioxide concentration in the air is another important factor for plant growth. The CO_2 concentration in the open air varies daily. On a sunny day the CO_2 level is fairly constant around 300 to 350 $\mu l/l$ due to photosynthesis by plants, and it rises to around 400 to 450 at night. The CO_2 level is more important under cover, especially in plastic houses where air exchange is normally restricted in order to raise temperatures in a cold climate. As a result, the carbon dioxide concentration falls below the outside level during the daytime due to photosynthetic CO_2 uptake. Therefore, sometimes CO_2 enrichment is conducted in plastic houses.

In CO_2 flow above the open ground in the daytime, the plant layer is a sink, and there are two sources of CO_2, one in the air layer above the plants and the other in the soil layer. In plastic houses, CO_2 is supplied from outside through ventilation instead of through flow from the air layer above the plants. CO_2 generation in soil is therefore an important source in both open fields and plastic house environments. CO_2 in soil is released through plant root respiration, microbial activity, chemical changes and physical phenomena.

7.2. CO_2 CONCENTRATION IN SOIL LAYER (CUC70)

There are not many reports on the CO_2 environment in soil, but Yabuki (1966) measured CO_2 concentration in soil and considered its daily and seasonal variation. In simulations by van Bavel (1951) and Ito (1979), steady-state patterns in several soil layers were demonstrated clearly, but periodic changes in CO_2 concentration were not considered. It is clear that non-linear models are needed in order to analyze the effects of soil temperature on CO_2 diffusion. Typical hourly changes and trends over three days are shown in Fig. 7.1.

Large changes at deeper soil layers are striking; it is apparent that temperature has an effect on the CO_2 concentration.

CO_2 diffusion in soil is similar to heat transfer as well as to water vapor transfer in soil. Therefore, let us consider the same model we used for soil temperature in Chapter 3. In this model, CO_2 concentration (C1) in the first soil layer shown in Fig. 3.10 is given as

$$IC1 = DS * (CCF - C1)*2.0 / Z + DS * (C2 - C1) / Z) / Z + gen1 \qquad (7.1)$$

where IC1 is the differential form of the C1 related equation, which is one of the equations to be solved simultaneously. DS is the diffusion coefficient (m^2/hr); CCF, C1 and C2 are CO_2 concentrations just above the ground, in the top layer, and in the second layer, respectively; Z is the depth of the soil layer; and **gen1** is CO_2 generation in the same soil layer $(m^3/m^3/hr)$ derived using the following equation:

$$\textbf{gen1} = \textbf{CCG1} * \textbf{fgen1}$$

where **CCG1** is the coefficient of CO_2 generation and **fgen1** is the amount of CO_2 generated in soil layer 1 by micro-organisms. It has also been found that the effect of temperature on CO_2 generation is much greater than that on the diffusion coefficient (**DS**), so **DS** is assumed to be constant in the program.

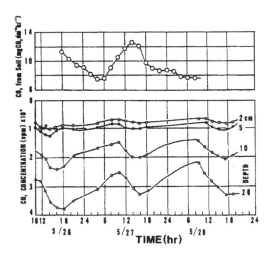

Figure 7.1. Hourly patterns of CO₂ concentration change in soil (after Yabuki, 1966).

Biological respiration is a **Q10** function of temperature; that is, for every 10°C increase in temperature, respiration is doubled (see Fig. 7.2). Therefore, it is assumed that CO_2 generation due to biological reactions is in this form, as indicated by the array of [**Xtemp, Yco2**] as shown in Fig. 7.3b and Fig. 7.6 (0 to 30°C). Non-dimensional generation is 0.0 at 0°C, 0.5 at 10°C, 1.0 at 20°C, 2.0 at 30°C, 1.5 at 40°C, 1.0 at 50°C and 0.0 at 60°C. Thus, the array of **Xtemp** is [0 10 20 30 40 50 60] and **Yco2** is [0.0 0.5 1.0 2.0 1.5 1.0 0.0]. A function for one-dimensional interpolation, **interp1**, is used to derive the amount of CO_2 generated according to the soil temperature.

$$\textbf{fgen1} = \textbf{interp1}(\textbf{Xtemp, Yco2, T1, 'nearest');}$$

The third parameter is the value of Xtemp for interpolation, and the fourth parameter is the method used for the interpolation. The graph of [**Xtemp, Yco2**] and several methods of interpolation can be found in Fig. 7.6.   Refer to section 7.5 of this chapter for more details.

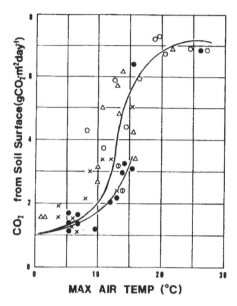

Figure 7.2. The relationship between maximum air temperature and CO2 release from soil surface (after Yabuki, 1966).

The main part of the program is more or less the same as that in Fig. 3.8.   Five equations **(IC1** to **IC5)** to find the CO$_2$ concentration in soil layers (**C1** to **C5)** have been added with boundary conditions, one of which is a **sin** function (**CCF)** to give CO$_2$ variation at the soil surface.

The value for **DS** has been changed from 0.002 to 0.02 (m^2/hr).   The former is close to the value van Bavel used in his simulation. When **DS** becomes larger, gas exchange is activated and the CO$_2$ concentration levels and amplitude of variation are decreased.   This is a much faster way to obtain a steady state, and a tendency to increase or decrease without reaching steady-state is also overcome.

CO$_2$ generation in the soil layers is assumed to be larger in the upper soil layers than in the lower layers: the CO$_2$ generation rate is 1×10^{-3} m^3/m^3/hr in the upper two layers and 0.7, 0.5 and 0.3×10^{-3} in the remaining three layers.   These figures are on the order of van Bavel's values (1951), and ten times smaller than those of Ito (1979).   As it is not easy to measure CO$_2$ generation in the soil layer, it is difficult to determine which order of magnitude is reasonable.   In the present simulation, these values are assumed in order to measure periodic changes similar to those of Yabuki (1966).   However, there are some data that indicate the CO$_2$ concentration in soil is up to ten times more than those simulated here.

A typical simulation result is shown in Fig. 7.4.   Since the value of **DS** in this
run is rather small (0.005), a tendency for CO_2 to slightly increase with periodic
change is apparent in deeper layers.   The amplitude of the CO_2 concentration in the
first layer is about 1000 $\mu l/l$ (0.1%), and the CO_2 concentration changes from 5000
to 6000 $\mu l/l$ (0.5 to 0.6%).   In the second layer, the amplitude is about 3000 $\mu l/l$
(0.3%), and the CO_2 concentration changes from 12000 to 15000 $\mu l/l$ (1.2 to 1.5%).
Amplitudes decrease to a degree similar to the experimental data in Fig. 7.1.

```
%    Carbon dioxide concentration in the soil layer              CUC70.m
%    Generation in each layer is assumed as a function of temperature
%    Function requires: co2insoil.m
%
function cuc70(out)
clc
disp('Please wait.');
global DS
if out==1, DS=0.005;end
if out==2, DS=0.02;end
tic
tstart=0;tfinal = 48;       % Define run time
y0 = [20 20 20 20 20 0.005 0.015 0.02 0.02 0.02]' ;   % initial conditions.
[t,y] = ode23t('co2insoil',[tstart tfinal],y0);
% calling function ode23t, constants & eqs are listed in 'co2insoil.m'
% Replace ode23t with ode45 for MATLAB version prior to 5.3
% [t,y]=ode45('co2insoil',[tstart tfinal],y0);
toc
if out==1
    h1=findobj('tag','Temperature');        close(h1);
    figure('tag','Temperature','Resize','on','MenuBar','none',...
    'Name','CUC70.m (Figure 1: Temperature in soil layers)',...
       'NumberTitle','off','Position',[140,140,520,420]);
    h=plot(t,y(:,1),t,y(:,2),t,y(:,3),t,y(:,4),t,y(:,5));
    set (h,'linewidth',2);
    axis([-inf,inf,10,40]);          grid on;
    xlabel('time elapsed, hr');    ylabel('Soil temperature, ^oC');
    legend('T1','T2','T3','T4','T5');
    title('Temperature in soil layer');
    figure('tag','CO2','Resize','on','MenuBar','none',...
    'Name','CUC70.m (Figure 2: CO2 concentration in soil layers)',...
       'NumberTitle','off','Position',[180,100,520,420]);
end
if out==2
    figure('tag','CO2','Resize','on','MenuBar','none',...
    'Name','CUC70.m (Figure 3: CO2 concentration in soil layers)',...
       'NumberTitle','off','Position',[220,60,520,420]);
end
h=plot(t,y(:,6),t,y(:,7),t,y(:,8),t,y(:,9),t,y(:,10));
set (h,'linewidth',2);
axis([0,48,0,0.03]);
grid on;
xlabel('time elapsed, hr');
ylabel('CO_2 cincentration, ND');
legend('C1','C2','C3','C4','C5');
co2title=['CO_2 concentration in soil layer when DS=' num2str(DS,5) ' m^2/hr'];
title(co2title);
disp('Thank you for using ');
disp(' ');
disp('CUC70: Porgram to calculate changes of soil temperature');
```

```
disp('        and CO2 concentration in soil layers with focus on DS value.');
disp(' ');
disp('You can enter ''close all'' in the command window to close figure windows.');
disp(' ');
```

Figure 7.3a. Main program to calculate CO_2 concentration in soil (**CUC70.m**).

```
% Subprogram to be used with cuc70 model                    co2insoil.m
function dy = co2insoil(t,y)
global DS
%[Xtemp Yco2] is the array of CO2 generated from soil;
Xtemp=[0 10 20 30 40 50 60];
Yco2=[0.0 0.5 1.0 2.0 1.5 1.0 0.0];
omega=2*pi/24;
%----------------------------------------PARAMETER--------------------
Z=0.1;          % Depth of each soil layer (m)
KS=5.5;CS=2000;
% KS=11;CS=2000;    % KS doubled
% KS=5.5;CS=4000;   % CS doubled
% KS: Soil thermal conductivity (kJ/m/C) and KS/3.6 (W/m/C)
% CS: Heat capacity of soil (kJ/m3/C)
%----------------------------------------CONSTANT---------------------
T0=25; TU=15; TBL=25;    % All in degree C
% T0: Avg. outside T, TU: Variation of T, TBL: Boundary soil T
%--------------------------------------------------------------------
clk=mod(t,24);
TF=T0+TU*sin(omega*(clk-8.));
% TF: Soil temp of surface layer (C),   t: time  in hr
% Maximum temp occurs at 2 o'clock in the afternoon
CCF=(0.04+0.01*sin(omega*(clk+6)))/100;
% CCF: CO2 concentration in the air
% DS=0.005;  % now a global variable
% DS: diffusion coefficient of CO2 in soil (m2/hr)
CCG1=0.001;CCG2=0.001;CCG3=0.0007;CCG4=0.0005;CCG5=0.0003;
% CCGx: CO2 generation coefficient in soil by microorganisms (m3/m3/hr)
%--------------------------------------------------------------------
T1=y(1); T2=y(2); T3=y(3); T4=y(4); T5=y(5);
C1=y(6); C2=y(7); C3=y(8); C4=y(9); C5=y(10);
IT1=(2*KS*(TF-T1)/Z+KS*(T2-T1)/Z)/Z/CS;
IT2=(KS*(T1-T2)/Z+KS*(T3-T2)/Z)/Z/CS;
IT3=(KS*(T2-T3)/Z+KS*(T4-T3)/Z)/Z/CS;
IT4=(KS*(T3-T4)/Z+KS*(T5-T4)/Z)/Z/CS;
IT5=(KS*(T4-T5)/Z+KS*(TBL-T5)*2/Z)/Z/CS;
%--------------------------------------------------------------------
fgen1=interp1(Xtemp,Yco2,T1,'nearest'); gen1=CCG1*fgen1;
IC1=(2*DS*(CCF-C1)/Z+DS*(C2-C1)/Z)/Z+gen1;
fgen2=interp1(Xtemp,Yco2,T2,'nearest');  gen2=CCG2*fgen2;
IC2=(DS*(C1-C2)/Z+DS*(C3-C2)/Z)/Z+gen2;
fgen3=interp1(Xtemp,Yco2,T3,'nearest');  gen3=CCG3*fgen3;
IC3=(DS*(C2-C3)/Z+DS*(C4-C3)/Z)/Z+gen3;
fgen4=interp1(Xtemp,Yco2,T4,'nearest');  gen4=CCG4*fgen4;
IC4=(DS*(C3-C4)/Z+DS*(C5-C4)/Z)/Z+gen4;
fgen5=interp1(Xtemp,Yco2,T5,'nearest');  gen5=CCG5*fgen5;
IC5=DS*(C4-C5)/Z/Z+gen5;
dy = [IT1 IT2 IT3 IT4 IT5 IC1 IC2 IC3 IC4 IC5]';
```

Figure 7.3b. Subprogram to calculate CO_2 concentration in soil (**CO2insoil.m**).

7.3. PROGRAM EXECUTION AND OUTPUT

Enter 'cuc70(1)' in the Command Window to execute the **cuc70** program. When variable **out** equals 1 (or 2), value 0.005 (or 0.02) will be assigned to variable **DS**. Fig. 7.4a and Fig. 7.4b show the outputs of 'cuc70(1)', and Fig. 7.4c shows the output of 'cuc70(2).

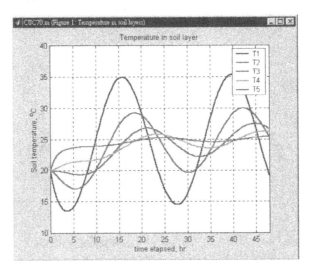

Figure 7.4a. Temperatures in soil layers when DS=0.005.

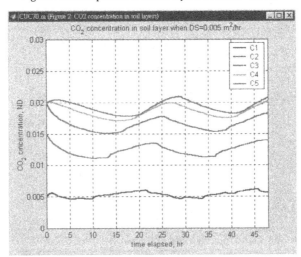

Figure 7.4b. CO_2 concentrations in soil layers when DS=0.005.

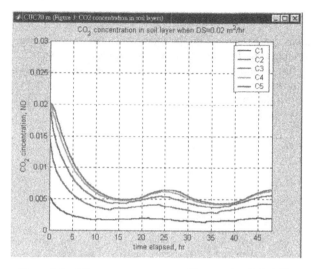

Figure 7.4c. CO₂ concentrations in soil layers when DS=0.02.

7.4. CO₂ CONCENTRATION IN A PLASTIC HOUSE AND VENTILATION

As mentioned in the introduction to this chapter, the CO_2 concentration in a plastic house is determined by the CO_2 balance in the house. There are two sources and one sink during the daytime. CO_2 brought in from the outside through ventilation and released from soil in the house are the sources, and plant photosynthesis is the sink.

Ventilation is usually defined as either a rate or an amount. Ventilation rate (**N**) is expressed in terms of the number of multiples of house volumes of air exchanged in an hour (1/hr), and ventilation amount (**Q**) is air volume per unit floor area per unit time ($m^3/m^2/min$).

Then, CO_2 balance in a house is given as

$$\mathbf{V * (dCI/dt) = V * N * (CO - CI) + AS * DS * SH *}$$
$$\mathbf{(CF - CI) - CONV * PH * AP} \tag{7.2}$$

where **V** is house volume (m^3); **CI**, **CO** and **CF** are CO_2 concentrations in the house, in the outside air and at the soil surface, respectively (m^3/m^3); **N** is ventilation rate (1/hr); **AS** is floor area (m^2); **DS** is diffusion coefficient of CO_2 (m^2/hr); **SH** is Sherwood number; **CONV** is the conversion factor from mg CO_2 to m^3; **PH** is photosynthesis rate (mg $CO_2/m^2/hr$); and **AP** is plant leaf area in a house (m^2). Using eq. 7.2, the model in Fig. 7.3 can be combined with the house model in Fig. 6.13, for example.

7.5. ONE-DIMENSIONAL INTERPOLATION

Several methods can be used to do the interpolation as shown in Fig. 7.5 and Fig. 7.6. The execution time follows the same order as listed in Fig. 7.5. The curves generated by the methods 'cubic' and 'spline' are much smoothes than those generated by the methods 'nearest' and 'linear'. If components of the vector X are at the same interval, the speed of execution can be improved by a leading '*' with a method such as '*linear'.

```
% Learning one-dimensional interpolation                    cuc70sup.m
%
x=0:10:60;
y=[0 0.5 1 2 1.5 1.0 0];
xi=0:1:60;
y1=interp1(x,y,xi,'nearest');
y2=interp1(x,y,xi,'linear');
y3=interp1(x,y,xi,'cubic');
y4=interp1(x,y,xi,'spline');
plot(x,y,'o',xi,y1,'g-',xi,y2,'r:',xi,y3,'c-.',xi,y4,'b--');
legend('Original','nearest','linear','cubic','spline');
```

Figure 7.5. Learning one dimensional interpolation.

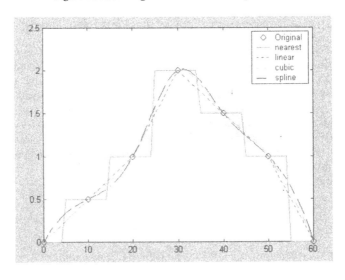

Figure 7.6. Output of program listed in Fig 7.5.

PROBLEMS

1.   Consider the reason why we have a rather irregular shaped curve for the change of CO$_2$ concentration in the first soil layer.

2.   Try to find how many days we need to get steady periodic change for both temperature and CO$_2$ concentration in all soil layers.

3.   Rerun the program **CUC70**, assuming that the diffusion coefficient of CO$_2$ **(DS)** is a function of temperature; that is, **DS** $= 0.005 + 0.001*$**TEMP**, where **TEMP** is soil temperature.

4.   Change the boundary condition for CO$_2$ from the CO$_2$ concentration at the soil surface to that in the air in the program **CUC70**.

5.   What is the effect of soil properties **KS** and **CS** on CO$_2$ concentrations in the soil layers?   Note that **KS** is the soil thermal conductivity (kJ/m/C) and **CS** is the heat capacity of the soil (kJ/m^3/C).   Rerun the program **CUC70** using the following 3 sets of parameter values:

        Case 1: KS $= 5.5$, CS=2000
        Case 2: KS $= 11$, CS=2000
        Case 3: KS $= 5.5$, CS=4000

CHAPTER 8

WATER AND WATER VAPOR ENVIRONMENT

8.1. INTRODUCTION

Water movement in soil is much more complicated than that of CO_2 because not only is water present in two phases, one liquid and the other gas, but also water movement is related to heat flow. However, in order to simplify the problem and understand the system clearly, it is again assumed that vapor flow in soil is independent of heat flow, as we assumed in the chapter on heat flow analysis. Furthermore, we will focus only on vapor movement in the present chapter, because in most cases water flow is assumed to occur in parallel with water vapor flow and they are combined. This can be condensed into what is the diffusion coefficient for both.

When soil is covered by vegetation, as is usual in agricultural fields, water is transported from the soil via roots, stems and leaves. This transport system in the form of a soil-plant-atmosphere continuum (SPAC) was introduced using water potential (Slatyer, 1963). It is convenient to use water potential to describe the water flow in SPAC. The concept of SPAC is clear, and sophisticated models for open fields have also been developed; the power of CSMP has also been demonstrated. Since evaporation from bare soil is the most important part of these rather complicated systems, a system without plants is considered in this chapter. In Chapter 10, the relationship between transpiration and environmental conditions is discussed in a physiological model of plant leaves.

Water potential, **SS** (J/kg), is proportional to the difference in chemical potential found by Gibb's free energy law and defined as

$$\mathbf{SS = (MYU - MYU0) / VOL = RC * AT / VOL * log(P / P0)} \qquad (8.1)$$

$$\mathbf{= RC * AT / GRV * log(P / P0)} \qquad (8.2)$$

where **MYU** and **MYU0** are chemical potentials of the water used in the study and pure water, respectively, **P** and **P0** are water vapor pressure and saturated water vapor pressure, respectively (Pa), **RC** is the gas constant (461.5 J/kg/K or 8.314 J/mole/K), **AT** is absolute temperature (T+273, K), **VOL** is mole volume of water (18.0 cm^3/mole), and **GRV** is acceleration of gravity (9.8 m/s^2) (*e.g.*, Koorevaar *et al.*, 1987). The unit of water potential is J/cm^3. This is dimensionally equivalent to pressure units such as N/cm^2, Pa and bar. It is also equivalent to the unit J/g because water density is 1 g/cm^3. As a practical unit, J/kg is used and is calculated by using 0.018 kg/mole for **VOL**. It is also shown that 1 J/kg equals 1 kPa. The bar is the most popular unit for pressure, and 1 bar is equal to 100 kPa. In eq. 8.1, the unit of water potential (**SS**) is in J/kg. In eq. 8.2, as the unit of water potential is kPa, **SS** is expressed in meters since one atmospheric pressure is approximately 10 m Aq and is equal to 100 kPa, and 1 kPa is equal to 0.1 m.

8.2. WATER AND WATER VAPOR MOVEMENT IN SOIL

Water movement in liquid form is considered first. Vertical water flow in soil is expressed as the following equation:

$$\mathbf{FLR} = -\ \mathbf{KT} * d\mathbf{SS}/d\mathbf{z} - \mathbf{KT} \tag{8.3}$$

where **FLR** is water flow rate ($m^3/m^2/hr$), **KT** is hydraulic conductivity (m/hr), **SS** is water potential (m), and **z** is soil depth (m).   The typical units for water potential are J/kg, kPa and m, as indicated in the introduction.   The conversion from the unit $m \cdot m/s^2$ to kPa is as follows: 1 $m \cdot m/s^2$ is equal to 0.1 m of pressure head of water, and 1 m of pressure head of water is 10 kPa.   Therefore, the unit $m \cdot m/s^2$ is equivalent to the unit J/kg.

Hydraulic conductivity is complicated and is known to be a function of the volumetric water content (**WW**, m^3/m^3).   In order to simplify the situation to show the general behavior of the model, some functional relationships from van Keulen (1975) are introduced.   Hydraulic conductivity (**KT**) is expressed as

$$\mathbf{KT} = \mathbf{AFGEN\_KTB(WW)} \tag{8.4}$$

and this relationship is depicted by a smooth curve in Fig. 8.1.

Water potential (**SS**) is a function of volumetric water content (**WW**) in the following way:

$$\mathbf{SS} = \mathbf{AFGEN\_STB(WW)} \tag{8.5}$$

Fig. 8.2 also shows the relationship between water potential and volumetric water content. In a similar way, vapor diffusivity (**DWV**, m^2/hr) is a function of volumetric water content (**WW**, m^3/m^3),

$$\mathbf{DWV} = \mathbf{AFGEN\_WVDT(WW)} \tag{8.6}$$

where **WW** is given in Fig. 8.1.

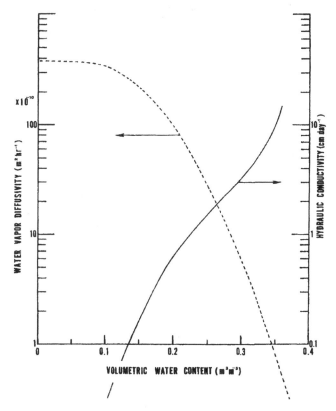

Figure 8.1. Hydraulic conductivity and water vapor diffusivity are given as functions of volumetric water content (after van Keulen, 1975).

Vapor flow is expressed as

$$\mathbf{WVF} = \mathbf{DWV} * d\mathbf{VPC} / d\mathbf{z} \qquad (8.7)$$

where **VPC** is vapor concentration (m³/m³).

Saturated water vapor pressure (**PWS**) was already given in the previous models as a function of temperature, and is expressed in a separate function definition:

$$[\mathbf{PWS}, \mathbf{WWW}] = \mathbf{FWSP(TTT)} \qquad (8.8)$$

Actual vapor pressure (**VPM**, mbar) is given as the product of **PWS** and **DF** (depression factor); furthermore, it is converted to vapor concentration (**VPC**, m³/m³).

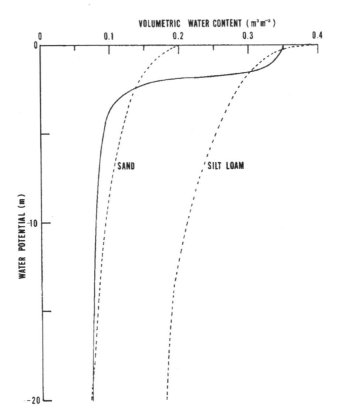

Figure 8.2. The relationship between water potential and volumetric water content; (Solid line after van Keulen, 1975; and dotted lines from Campbell, 1985).

8.3. WATER AND HEAT BALANCE IN SOIL LAYER (**CUC90**)

The complete model is given in Fig. 8.3, and the simulated results are given in Fig. 8.4. The main structure of the model is the same as that for the heat flow model (**CUC03**) discussed in Chapter 4. The water and water vapor flow model of van Keulen (1975) has been modified and added to the original heat flow model.

Evaporation was already taken into account in the previous model; conversion from water content to humidity ratio at the soil surface is still needed. First, using eq. 8.2, the depression factor (**DF**) is calculated as

$$\mathbf{DF = P / P0 = exp(SS * GRV / (RC * AT))} \qquad (8.9)$$

The humidity ratio at the soil surface (**WF**, kg/kg DA) is

$$WF = 0.622 * VPMF / (1.0E5 - VPMF) \qquad (8.10)$$

where **VPMF** is vapor pressure and is calculated by **PWSF*DF**, and **PWSF** is saturated vapor pressure (Pa).

The relationships listed in eqs. 8.1 - 8.10 are introduced into the model of **CUC03**. The basic structure of the model consists of five differential equations for temperature (**TF** – **T4**) and five equations for volumetric water content (**WWF** – **WW4**). Analogous to heat balance, water balance is given as

$$d\text{WW} / d\text{t} = (\text{FLR} + \text{WVF}) / d\text{z} \qquad (8.11)$$

The water balance expression for the surface layer is as follows:

$$\text{I\_WW}_F = (\text{FLR}_F - \text{FLR}_1 + \text{WVF}_F - \text{WVF}_1) / \text{Z0} \qquad (8.12)$$

where **WWF** is the water content in the surface layer, and it is the integration of water flow gradient (**FLR$_F$ - FLR$_1$**) / **Z0** and vapor flow gradient (**WVF$_F$ - WVF$_1$**) / **Z0**. The suffix **F** is for the surface layer, and the suffix 1 is for the first layer under the surface layer.

Hydraulic conductivity (**KT**) is a function of water content; thus it would be better to take the average of two adjacent layers. For example, **KT$_1$** is calculated as follows :

$$\begin{aligned}(\text{WW}_F * \text{AFGEN\_KTB}(\text{WW}_F) + \text{WW}_1 * \\ \text{AFGEN\_KTB}(\text{WW}_1)) / (\text{WW}_F + \text{WW}_1)\end{aligned} \qquad (8.13)$$

In a similar way, water vapor diffusivity of the first layer (**DWV$_1$**) is averaged:

$$\begin{aligned}(\text{Z0} * \text{AFGEN\_WVDT}(\text{WW}_F) + \text{Z1} * \\ \text{AFGEN\_WVDT}(\text{WW}_1)) / (\text{Z0} + \text{Z1})\end{aligned} \qquad (8.14)$$

During the execution, a progress report will be displayed in the Command Window showing 'Progress: xx.xx out of 48 hours' which means it is now at the xx.xx simulated time out of the total 48 hours simulated times. Due to the complexity of the model (solving 10 simultaneous differential equations) which can involve rapid changes of the values, execution with large calculation intervals may not be able to converge, thus leading to a long execution time for this model. The execution time is the longest among all models in this book.

An improvement was made in the program to accelerate the execution of the program by using options= **odeset('RelTol',0.005)** and **ode15s(.....,** options) functions. The **odeset** function allows users to create/alter the **ode** options structure and the 'RelTol' stands for relative error tolerance. The estimated error in each integration step satisfies e(i) <= max(**RelTol*abs(y(i))**, **AbsTol**(i)), where, 'y(i)' is the i th unknown variable involved in solving the **ode** function, and 'AbsTol' stands for absolute error tolerance and defaults to 1e-6. The default value of 'RelTol' of the **ode** function in **MATLAB** is 0.001. The larger the '**RelTol**' value, the shorter the execution time required. The current value of 0.005 was decided based on trial and error. The model will not converge if the value is increased to 0.01.

Two figures will be generated by the **CUC90** model. The first figure (not shown) displays the temperature regimes in the soil layers and is similar to the output of the **CUC03** model. The second figure, shown in Fig. 8.4, is the water content in the soil layers. As shown in the figure, the effect of the initial condition remains at first, but soon the water content drops in all soil layers. Water content begins to increase in the early part of the morning because the humidity ratio of the air is greater than that in the soil surface layer and water vapor flows from the air to the soil surface. After sunrise, the water content in all soil layers drops rapidly, and most dramatically in the upper layers. Water content is recovered in the nighttime mainly due to condensation at the soil surface, as indicated in Fig. 8.4. Dew-point temperature at night is higher than the soil surface temperature, and this causes condensation. The soil surface does not recover to the original water content level, however, and the water content in each soil layer gradually decreases with the passage of time.

```
% Program to calculate temperature                               CUC90.m
%          and water regimes in the soil layer
% function required: soil90.m
%
clear all; clc
t0=0;tfinal=48;
y0=[10;10;10;10;10;0.3;0.31;0.33;0.34;0.35];
%    TF T1 T2 T3 T4 WWF  WW1  WW2  WW3  WW4
options = odeset('RelTol',0.005);
[t,y]=ode15s('soil90',[t0 tfinal],y0,options);
%-----------------------------------------------------------------------
h1=findobj('tag','cuc90_part1');    close(h1);
figure('tag','cuc90_part1','Resize','on','MenuBar','none',...
    'Name','CUC90.m (Figure 1: Soil temperature regimes in soil layers)',...
    'NumberTitle','off','Position',[160,80,520,420]);
plot (t,y(:,1),'r+-',t,y(:,2),'b^:',t,y(:,3),'k*-',t,y(:,4),'ko-',...
    t,y(:,5),'k-','linewidth',2);
grid on;
xlabel('time elapsed, hr');    ylabel('Temperature, ^oC');
legend('T_f','T_1','T_2','T_3','T_4',2);
%-----------------------------------------------------------------------
h1=findobj('tag','cuc90_part2');    close(h1);
figure('tag','cuc90_part2','Resize','on','MenuBar','none',...
    'Name','CUC90.m (Figure 2: Water contents regimes in soil layers)',...
    'NumberTitle','off','Position',[200,40,520,420]);
plot (t,y(:,6),'r+-',t,y(:,7),'b^:',t,y(:,8),'k*-',t,y(:,9),'ko-',...
    t,y(:,10),'k-','linewidth',2);
grid on;
legend('WWF','WW1','WW2','WW3','WW4',1);
xlabel('time elapsed, hr');    ylabel('Water content, m^3/m^3');
fprintf('\n\n');
disp(' You can enter ''close all'' to close figure windows.');
```

Figure 8.3a. Main program to calculate water regime in soil (**CUC90.m**).

```
% Functions used in CUC90 model.                          soil90.m
% Including:
%    soil90, tabs, solar, WPCH, FWSP, AFGEN_KTB, AFGEN_STB, AFGEN_WVDT
%
function dy = soil90(t,y)
Tavg=10.0; TU=5.0; TBL=10.0; TD=4.5;RP=2000;
KS = 5.5; CS= 2.0E+3; RHO=1.164;QH=36;HS= 25.2;SIG = 20.4;
Z0=0.01; Z1=0.05; Z2=0.1; Z3=0.2; Z4=0.1;
ALF=0.7; EPSF=0.95; HLG=2501.0;LE=0.9;
KM=HS/LE;
[PD,WO]=FWSP(TD);
EPSA=0.711+(TD/100)*(0.56+0.73*(TD/100));
%-----------------------------------------------------------------
%        Brief Description of constants
%-----------------------------------------------------------------
% RHO:   Density of air (kg/m^3)
% TD:    Outside dew point temperature, in degree C
% KS:    in(kJ/m/C/hr) and KS/3.6 (W/m/C) also CS (kJ/m^3/C)
% QH:    ventilation air flow rate (m^3/m^2/hr), 8 cfm/ft^2 = 146 m/hr
% HS:    Convective heat transfer coefficient
% SIG:   Stefan-Boltzmann constant (kJ/m^2/K^4/hr) = 5.67(W/m^2/K^4)
% Z0-Z4: Depths of soil layer (m)
% ALF:   Absorptivity of solar radiation at soil surface
% EPSF:  Emissivity of soil surface
% HLG:   Latent heat
% EPSA:  Emissivity of air layer
%-----------------------------------------------------------------
%        Brief Description of parameters
%-----------------------------------------------------------------
% DFF,   DF1,...,DF5 :           Depression Factor
% VPMF,  VPM1,..,VPM5:           Vapor Pressure, in Pa
% PWSF,  PWS1,..,PWS5:           Saturated Vapor Pressure, in Pa
% PWSMF, PWSM1,.,PWSM5:          Saturated Vapor Pressure, in mbar
% WWWF,  WWW1,..,WWW5:           Absolute humidity
% SSF,   SS1,...,SS5:            Water potential
% TF,    T1,.....,T5:            Temperature, in degree C
% FLRF,  FLR1,..,FLR5:           Water flow rate
% WVFF,  WVF1,..,WVF5:           Vapor flow rate
% WWF,   WW1,...,WW5:            Water content
% VPCF,  VPC1,..,VPC5:           Water vapor concentration
% DWV1,    ,..,DWV5:             Water vapor diffusivity, in m^2/hr
% Function of water content (WW1-5) and is averaged.
%-----------------------------------------------------------------
clc;
fprintf('Progress: %f out of 48 hrs\n',t);    % Display  the  execution
progress
clk = mod(t,24);                      % Time (hr)
OMEGA=2*pi/24;                        % converter
TO= Tavg+TU*sin(OMEGA*(clk-8));       % TO is outdoor air temperature
RAD=solar(RP,OMEGA,clk);             % Calling function solar()
%-------------------------------------------[Surface layer]------
TF=y(1); T1=y(2); T2=y(3); T3=y(4); T4=y(5);
WWF=y(6); WW1=y(7); WW2=y(8); WW3=y(9); WW4=y(10);
[PWSF,WWWF]=FWSP(TF);
TERF=WWF*AFGEN_KTB(WWF)/2400;
TER1=WW1*AFGEN_KTB(WW1)/2400;
% AFGEN_KTB: Hydraulic conductivity KTB (cm/day)
%    is a function of water content (WWF-5)
%    and is averaged between adjacent two layers.
% Water flow is a function of water potential difference (SSF-5) and gravity.
SSF=-1/100*AFGEN_STB(WWF);
```

```
DFF=WPCH(TF,SSF);
VPMF=PWSF*DFF; % DFF is ratio of vapor pressure and saturated vapor pressure
WF=WWWF*DFF;
% Assuming DFF equals ratio of absolute humidity and saturated absolute
humidity
%WF=0.622*VPMF/(1.e5-VPMF);    another way to calculate WF (this is eq.8.10)
TO4=tabs(TO);  % calling function TABS()
TF4=tabs(TF);
FLRF=0;                % assuming no irrigation
SS1=-1/100*AFGEN_STB(WW1);
FLR1=(TERF+TER1)/(WWF+WW1)*((SSF-SS1)*2/(Z0+Z1)+1);
VPCF=1.3411/1000/(273+TF)*VPMF;
[PWS1,WWW1]=FWSP(T1);
PWSM1=PWS1*0.01;
DF1=WPCH(T1,SS1);
VPM1=PWSM1*DF1;
VPC1=1.3411/1000/(273+T1)*VPM1;
DWV1=(Z0*AFGEN_WVDT(WWF)+Z1*AFGEN_WVDT(WW1))/(Z0+Z1)*3.6;
I_TF=(ALF*RAD+EPSF*SIG*(EPSA*TO4-TF4)+HS*(TO-TF)+HLG*KM*(WO-WF)+...
    KS*(T1-TF)*2/(Z0+Z1))/CS/Z0;                        % 1st eq.
WVFF=-KM*(WF-WO);
WVF1=DWV1*(VPCF-VPC1)*2/(Z0+Z1);
I_WWF=(FLRF-FLR1+WVFF-WVF1)/Z0;                         % 2nd eq.
%------------------------------------------------------[First layer]----
I_T1=(KS*(TF-T1)*2/(Z0+Z1)+KS*(T2-T1)*2/(Z1+Z2))/CS/Z1;  % 3rd eq.
SS2=-1/100*AFGEN_STB(WW2);
TER2=WW2*AFGEN_KTB(WW2)/2400;
FLR2=(TER1+TER2)/(WW1+WW2)*((SS1-SS2)*2/(Z1+Z2)+1);
[PWS2,WWW2]=FWSP(T2);
PWSM2=PWS2/100;
DF2=WPCH(T2,SS2);
VPM2=PWSM2*DF2;
VPC2=1.3411/1000/(273+T2)*VPM2;
DWV2=(Z1*AFGEN_WVDT(WW1)+Z2*AFGEN_WVDT(WW2))/(Z1+Z2)*3.6;
WVF2=DWV2*(VPC1-VPC2)*2/(Z1+Z2);
I_WW1=(FLR1-FLR2+WVF1-WVF2)/Z1;                         % 4th eq.
%-----------------------------------------------------[2nd layer]-------
I_T2=(KS*(T1-T2)*2/(Z1+Z2)+KS*(T3-T2)*2/(Z2+Z3))/CS/Z2;  % 5th eq.
SS3=-1/100*AFGEN_STB(WW3);
TER3=WW3*AFGEN_KTB(WW3)/2400;
FLR3=(TER2+TER3)/(WW2+WW3)*((SS2-SS3)*2/(Z2+Z3)+1);
[PWS3,WWW3]=FWSP(T3);
PWSM3=PWS3/100;
DF3=WPCH(T3,SS3);
VPM3=PWSM3*DF3;
VPC3=1.3411/1000/(273+T3)*VPM3;
DWV3=(Z2*AFGEN_WVDT(WW2)+Z3*AFGEN_WVDT(WW3))/(Z2+Z3)*3.6;
WVF3=DWV3*(VPC2-VPC3)*2/(Z2+Z3);
I_WW2=(FLR2-FLR3+WVF2-WVF3)/Z2;                         % 6th eq.
%-----------------------------------------------------[3rd layer]-------
I_T3=(KS*(T2-T3)*2/(Z2+Z3)+KS*(T4-T3)*2/(Z3+Z4))/CS/Z3;  % 7th eq.
SS4=-1/100*AFGEN_STB(WW4);
TER4=WW4*AFGEN_KTB(WW4)/2400;
FLR4=(TER3+TER4)/(WW3+WW4)*((SS3-SS4)*2/(Z3+Z4)+1);
[PWS4,WWW4]=FWSP(T4);
PWSM4=PWS4/100;
DF4=WPCH(T4,SS4);
VPM4=PWSM4*DF4;
VPC4=1.3411/1000/(273+T4)*VPM4;
DWV4=(Z3*AFGEN_WVDT(WW3)+Z4*AFGEN_WVDT(WW4))/(Z3+Z4)*3.6;
WVF4=DWV4*(VPC3-VPC4)*2/(Z3+Z4);
```

```
I_WW3=(FLR3-FLR4+WVF3-WVF4)/Z3;                          % 8th eq.
%--------------------------------------------------[4th layer]-------
I_T4=(KS*(T3-T4)*2/(Z3+Z4)+KS*(TBL-T4)*2/Z4)/CS/Z4;      % 9th eq.
SS5=0;
TER4=WW4*AFGEN_KTB(WW4)/2400;
WSAT=0.35;
KSAT=AFGEN_KTB(WSAT)/2400;
KA5=(WW4*AFGEN_KTB(WW4)/2400+KSAT*WSAT)/(WW4+WSAT);
WTB=-1; % Assumed there is no water table.
% parameter for water table, if positive there is water table
%                      and if negative FLR5=0
if WTB<0;
   FLR5=0;
else
   FLR5=KA5*(SS4-SS5)*2/Z4;
end
WVF5=0;
I_WW4=(FLR4-FLR5+WVF4-WVF5)/Z4;                          % 10th eq.
%-------------------------------------------------------------------
dy=[I_TF;I_T1;I_T2;I_T3;I_T4;I_WWF;I_WW1;I_WW2;I_WW3;I_WW4];
return
%-------------------------------------------------------------------
function ktb_value=AFGEN_KTB(WW)
xtemp=[0.001 0.01 0.05 0.075 0.1 0.125 0.15 0.175 0.2 0.225 0.25 0.275 0.3 0.325
0.35];
ytemp=[1.e-8 1.e-7 2.5e-6 5e-5 5e-3 0.07 0.15 0.26 0.65 1 1.5 2.2 3.2 5
10];
ktb_value=interp1(xtemp,ytemp,WW,'linear');
%-------------------------------------------------------------------
function stb_value=AFGEN_STB(WW)
xtemp=[0.001 0.01 0.025 0.05 0.075 0.1 0.125 0.15 0.175 0.2 0.225 0.25 0.275 0.3 0.325 0.3325 0.35 1.35];
ytemp=[1.e9 1.e7 2.5e6 2.e5 1.475e3 365 265 230 215 190 182 170 158 151 126 60 0
-600];
stb_value=interp1(xtemp,ytemp,WW,'linear');
%-------------------------------------------------------------------
function wvdt_value=AFGEN_WVDT(WW)
xtemp=[0.01 0.05 0.1 0.2 0.35 0.4];
ytemp=[2.89e-8 2.55e-8 2.46e-8 9.84e-9 8.68e-11 1.e-11];
wvdt_value=interp1(xtemp,ytemp,WW,'linear');
%-------------------------------------------------------------------
function DF=WPCH(TTT,SSS)
GRV=9.8;    % Gravity (m/s^2)
RC=461.5;   % Gas constant (J/kg/k) and m^2/s^2=J/kg
AAA=SSS*GRV/(RC*(TTT+273));
DF=exp(AAA);
%-------------------------------------------------------------------
function [PWS, WWW]=FWSP(TTT)
Patm=101325; % Patm: assuming at sea level, in Pa, 1 atmosphere is 0.1
MPa
TQQ=TTT+273.16;
T10=TQQ/100.0;
if TTT>0
A=-5800.2206/TQQ+1.3914993-0.04860239*TQQ;
B=0.41764768*T10*T10-0.014452093*T10*T10*T10;
C=6.5459673*log(TQQ);
else
A=-5674.5359/TQQ+6.3925247-0.9677843*T10;
B=0.62215701E-2*T10*T10+0.20747825E-2*T10*T10*T10;
C=-0.9484024E-4*T10*T10*T10*T10+4.1635019*log(TQQ);
end
BETA=A + B + C;
```

```
PWS=exp(BETA);
WWW=0.622*PWS/(Patm-PWS);
%------------------------------------------------------------
function TT4 = tabs(TT)
    TAA = (TT+273.16)/100.0;          TT4 = TAA*TAA*TAA*TAA;
%------------------------------------------------------------
function rad = solar(RP,OMEGA,clk)
%     Calculation of solar radiation: RAD
%     Sunrise is 6 o'clock
    rad= RP*sin(OMEGA*(clk-6.0));
    if rad<=0,    rad=0;      end
```

*Figure 8.3b. Functions used in **CUC90** model (**soil90.m**).*

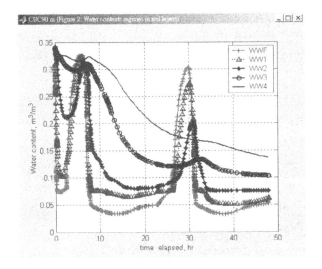

Figure 8.4. Water content in soil layers.

8.4. INTERACTION BETWEEN WATER MOVEMENT AND HEAT FLOW

Interactions of heat and moisture flow have been investigated in many researches. It was concluded that where temperature gradients are important, such as near the soil surface, simultaneous heat and mass transfer analyses gives a better fit with the experimental data than the isothermal diffusion equation (see Hillel, 1980). Simultaneous equations for analyzing non-isothermal transfer of vapor and liquid water under combined temperature and moisture gradients are offered by Philip and de Vries (in Hillel, 1980) in the following way:

$$\mathbf{FLR + WVF = -DWD} * d\mathbf{T}/d\mathbf{z} - (\mathbf{KT + DWV}) * d\mathbf{WW} / d\mathbf{z} - \mathbf{KT} \qquad (8.15)$$

$$\mathbf{HFL = -KS} * d\mathbf{T}/d\mathbf{z} - \mathbf{HLG} * \mathbf{DWV} * d\mathbf{WW} / d\mathbf{z} \qquad (8.16)$$

where the downward flow is taken as positive in both equations and newly introduced symbols are thermal water diffusivity **DWD**, and heat flow rate **HFL**. The analysis is beyond the scope of the present book.

Another mass flow in the soil layer is salt or ion movement. Ion movement is not covered in the present book, but it would not be difficult to expand the models in this book for these purposes.

MATLAB FUNCTIONS USED

---

odeset   Create/alter ODE OPTIONS structure. With no input arguments, **odeset** displays all property names and their possible values.

OPTIONS = ODESET('NAME1',VALUE1,'NAME2',VALUE2,...) creates an integrator options structure OPTIONS in which the named properties have the specified values. Any unspecified properties have default values. The properties involved in this chapter are as follows:

- RelTol - Relative error tolerance. This scalar applies to all components of the solution vector, and defaults to 1e-3 (0.1% accuracy) in all solvers. The estimated error in each integration step satisfies

    $e(i) <= \max (RelTol*abs(y(i)),AbsTol(i))$.
- AbsTol - Absolute error tolerance. A scalar tolerance applies to all components of the solution vector. Elements of a vector of tolerances apply to corresponding components of the solution vector. AbsTol defaults to 1e-6 in all solvers.

PROBLEMS

1. Calculate the water potential of the air if the relative humidity is 75%, and compare order of the magnitude with that of soil under normal conditions.

2. Explain the physical meaning of the statement for calculating **FLR5** in the program of **CUC90**.

3. Explain why eq. 8.13 uses **WW** (water content) as the weighing factor and eq. 8.14 uses **Z** (depth of the soil layer) as the weighing factor?

4. **INTERP1** is the function of 1-D interpolation of **MATLAB**. There are several methods available in this function, such as 'linear', 'nearest', 'cubic', 'spline' and 'pchip'. Use **function ktb_value=AFGEN_KTB(WW)**, listed in Fig. 8.3b, as an example, and write a **MATLAB** program to plot the differences of these methods.

5. **DFF** is the ratio of vapor pressure to saturated vapor pressure. As listed in Fig. 8.3b, **DFF** is used to calculate **WF** using the scripts of '**WF = WWWF * DFF;**' and assuming **DFF** is also the ratio of absolute humidity (kg/kg DA) of vapor to absolute humidity of saturated vapor. Is there any difference if we change the script to '**WF = 0.622 * VPMF / (1.e5 - VPMF);**'?

CHAPTER 9

CONTROL FUNCTION

9.1. INTRODUCTION

Models of greenhouses take various forms other than just greenhouses with single or double covering layers. Modelling of greenhouses has been studied extensively (Takakura, 1989). Static models are still predominant, and they are powerful for analyzing particular problems such as comparison of the effectiveness of various thermal screens and estimation of total heat requirements based on actual measurements.

In order to investigate the effectiveness of reflective blinds and the Fresnel prism effect of coverings, light penetration models have been developed.

Several models of total and dynamic systems have been developed. New innovations such as expert systems and practical feed-forward control systems will be included in the models, although there are several problems to be solved such as determination of heat transfer coefficients and sky temperature.

Greenhouse models consist of four sub-systems: 1) light penetration, 2) heat and mass transfer, 3) control function and 4) plant growth. The models are classified as either static or dynamic and as either total- or sub-systems (see Fig. 9.1). Static analyses using sub-systems are predominant for analyzing particular problems such as comparing the effectiveness of various thermal screens. In order to estimate overall heat requirements based on measurements in practical greenhouses, static analyses using total systems have been used. The light environment is one of the most important sub-systems for plant growth, and can be treated separately. Therefore, the light environment has been studied independently, and its models can be either static or dynamic. This sub-system can be included in the total system, of course.

Total and dynamic system models are defined as models that can predict inside environmental factors in the dynamic sense.

The sub-model for plant growth is one of the most difficult to include in the total system, but it is essential. Many primitive models have been reported. Rather sophisticated plant growth models with expert systems are involved in a new area of study although some difficulties have been noted.

Protected cultivation is now being investigated in two main ways. In one, we looked at environmental control using films in static ways; the other involves dynamic mechanical systems such as heaters and coolers. Where the outside climate is mild, most greenhouses are not heated. Even in Japan, more than 65% of the total covered cultivation area is not heated. In Mediterranean countries, there are a large number of unheated greenhouses. On the other hand, in northern Europe, greenhouses with heating systems are essential.

Several control functions are necessary for a heating system, and they must be considered as a unit. For example, in most cases air temperature and humidity interact with each other.   It is possible to change air temperature without changing absolute humidity, but relative humidity changes as a result.   A change in inside air temperature by ventilation changes the CO_2 concentration in the greenhouse. Therefore, it is rather difficult to control one factor without affecting others.

However, if we focus on control machines such as heaters or coolers and the main effect of these machines is on one environmental factor such as air temperature, the situation is simplified and can be a good example of actual and complicated control functions.   In the present chapter, the relationship between one control machine and one environmental factor is considered for two typical control logics; one is very popular and well-known feedback, and the other is feedforward.

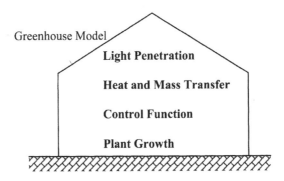

Figure 9.1. Greenhouse sub-models.

9.2. SYSTEM RESPONSE

Let us consider a simple example.   Suppose there is a container of water boiling and we put an egg into it; how will the egg's temperature change?   It will rise from the room temperature to the boiling temperature.   In fact, the temperature distribution varies from one part of the egg to another, and phase change occurs. But in the present example, it is assumed that the temperature distribution is uniform and there is no phase change.

If we generalize the temperature change of the egg, input it to the control system, and assume the temperature surrounding the egg changes from 0 to 1, then the response curve of the system (egg temperature) is similar to that shown by the solid line in Fig. 9.2(a).   This is called the system response with first-order delay.   The basic equation of the system response with first-order delay to the unit step input is given as

$$\textbf{RESP} = 1 - \textbf{exp}(-\textbf{t} / \textbf{DELA})\qquad(9.1)$$

This is a solution to the following differential equation:

$$\mathbf{DELA}*dy/dt + \mathbf{y(t)} = \mathbf{EE(t)} \qquad (9.2)$$

with a constant, that is $\mathbf{EE(t)} = 1$.

If the right-hand side of the equation is in the general form, $\mathbf{X(t)}$, then the general solution to eq. 9.2 is

$$\mathbf{y(t) = exp(-t/DELA)/DELA * \left(\int exp(t/DELA) * X(t)\,} dt + \mathbf{C)} \qquad (9.3)$$

If the heat content of the egg is large, its temperature does not rise immediately. A system response of this kind is called a system response with second or higher-order delay. The shape of this type of response is indicated by the chain curve in Fig. 9.2(a). These response curves are simplified and approximated as the response with first-order delay and dead time, as shown by the dotted line in Fig. 9.2(a). This can be interpreted as the response to a unit step change $\mathbf{RESD(t)}$ which is shown in Fig. 9.2(b).

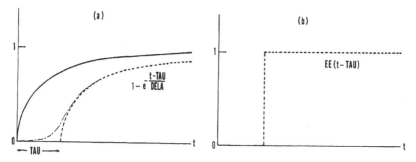

Figure 9.2. Response curve of the system.

This curve is expressed by $\mathbf{RESF(t)}$ as

$$\mathbf{RESF(t) = (1.0 - exp(-(t - TAU)/DELA))} \qquad (9.4)$$

and

$$\mathbf{RESD(t) = EE(t - TAU)} \qquad (9.5)$$

where t is time (min), $\mathbf{TAU}$ is dead time (min or hour) and $\mathbf{DELA}$ is delay time (min or hour). The function $\mathbf{EE(t)}$ is a unit step function which changes from 0 to 1 at $\mathbf{TIME}$ 0 and remains as 1. Therefore, the function $\mathbf{RESD}$ is the unit step function, which is delayed by $\mathbf{TAU}$.

In $\mathbf{CSMP}$ on mainframes, a $\mathbf{CSMP}$ function which is called $\mathbf{REALPL}$ is available. This is the function used to calculate eq. 9.3. Modelling can be simplified with this function, but it is not available in $\mathbf{micro\text{-}CSMP}$ and $\mathbf{MATLAB}$.

9.3. PID CONTROL

The most common type of feedback control logic is PID (Proportional, Integral and Differential) control. The error (**ER**) is defined as the difference between a set point (**SP**) and a controlled value (**CV**):

$$ER = SP - CV \qquad (9.6)$$

Therefore, the process variable (**PV**) is given as

$$PV = KK*(ER + \frac{1}{TL} \int_o^t ER * dt + TD * dER/dt) \qquad (9.7)$$

where **KK** is a proportionality constant, **TL** is a time constant for integration and **TD** is a time constant for differentiation.

9.4. TEMPERATURE CONTROL LOGIC (CUC120)

Let's consider the air temperature control of a floor-heating greenhouse, which is shown in Fig. 9.3. This system has a large heat mass in the floor, which creates a large time lag in the response. The system response is given by eq. 9.3 for an arbitrary input and by eq. 9.4 for a unit step input. If we use a PID controller to control the system, the feedback control logic is given by eqs. 9.6 and 9.7, where the unit for **CV** and **KK** is °C and the unit for **TL** and **TD** is hr. In CSMP, a step function is available; that is,

$$Y = STEP(TAU) \qquad (9.8)$$

which means that **Y** is 0 for **t** less than **TAU**, and **Y** is 1 for t equal to or greater than **TAU** as shown in the following **MATLAB** script.

```
If  (t < TAU)
     Y=0;
else
     Y=1;
end
```

In eq. 9.7, the derivative term can be approximated using numerical differentiation, *i.e.*, difference of **ER** over difference of **t**. The integral term can also be approximated using summation as shown in Fig. 9.4. **MATLAB** provides a function for integration that will be introduced in the next program. Normally, process variables (such as **PV** in this case) are used for control facilities such as heaters and coolers. Their units are, for example, expressed by voltage and valve positions. In order to simplify the problem, the unit of **PV** in the present case is assumed to be the same as that for the controlled value -- that is, temperature. The change in **PV** is in turn the input to the system -- that is, **X(t)** in eq. 9.3. Then, delay is introduced by using the source code listed in the % ---[Delay] section as shown in Fig. 9.4.

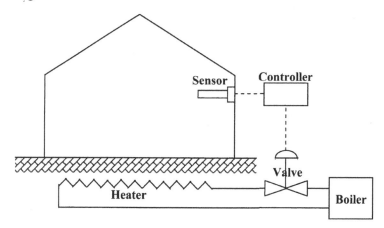

Figure 9.3. A floor-heating greenhouse.

Users can enter 'cuc120', 'cuc120(1)',.., 'cuc120(5)' in the Command Window to run this program. Typical results are shown in Figs. 9.5a, b and c. The response to the unit step change of the set point from 0 to 1°C, which is shown by the line with plus signs, has a 3 hour delay and increases rapidly. In Fig. 9.5a, the response first overshoots the set point and then reaches close to the set point in 48 hours. This means it takes two days to reach steady state after a unit step change. The line with star signs is the error curve, and it is clear that the error decreases rapidly with time. This is not far from reality. It is not difficult to imagine how the situation will be if the input changes periodically. The value **KK** is not only the proportionality constant in PID control logic but also a constant to convert the process variable to the controlled value including the process gain. Various shapes of the response can be obtained by changing the value of **KK** and/or **TAU**. If **KK** is large, overshoot appears (see Fig. 9.5a). Undershoot results from a smaller value of **KK** (see Fig. 9.5b). Smaller oscillations around the set point can be expected after 48 hours, when **TAU** is larger and **KK** is smaller as shown in Fig. 9.5c.

```
% Program for PID control,                                        cuc120.m
function cuc120(pid)
if nargin==0                      % if number of arguments is 0,
   pid=2;                         % case 2 is the default
end
if pid==1                         % case 1: PID control
   Tot_hrs=3600; logic='PID';
   kk = 0.01;tau = 120.0;         % small kk and late response (long delay)
elseif pid==2                     % case 2: PID control
   Tot_hrs=48; logic='PID'; kk=0.1;tau=3.0;% medium kk causing overshoot
elseif pid==3                     % case 3: PID control
   Tot_hrs=48; logic='PID'; kk=0.05;tau=3.0;% small kk causing undershoot
elseif pid==4                     % case 4: PI control
   Tot_hrs=48; logic='PI'; kk = 0.1;tau = 3.0;
elseif pid==5                     % case 5: PD control
   Tot_hrs=48; logic='PD'; kk = 0.1;       tau = 3.0;
end
tl = 0.9;   td = 0.2;   dela = 5.0;
sp = ones(1, Tot_hrs+1);  pv = zeros(1, Tot_hrs+1);
cv=pv;      er=pv;
er(1, 1) = sp(1, 1) - cv(1, 1);   er(1, 2) = sp(1, 2) - cv(1, 2);
total_er = er(1, 1) + er(1, 2);   total_2 = 0;   dt=1;
for t = 2: 1: Tot_hrs;
   switch logic
   case 'PID'                                         % PID control logic
      pv(1,t+1)=kk*(er(1,t)+td*(er(1,t)-er(1,t-1))/dt+1.0/tl*total_er);
   case 'PI'                                          % PI control
      pv(1, t+1)=kk*(er(1, t)+1.0/tl*total_er);
   case 'PD'                                          % PD control
      pv(1, t+1)=kk*(er(1, t)+td*(er(1, t)-er(1, t-1))/dt);
   case 'P'                                           % P control
      pv(1, t+1)=kk*er(1, t) ;
   case 'I'                                           % I control
      pv(1, t+1)=kk*(1.0 / tl * total_er);
   case 'D'                                           % D control
      pv(1, t+1)=kk*(td*(er(1, t)-er(1, t-1))/dt);
   end
%-------[Delay]-------------------------------------------------------
   if( t < tau )
      cv(1, t+1) = 0;
   else
      tmp = t-tau;      cv(1, t+1) = exp(-tmp/dela) / dela * total_2;
      total_2 = total_2 + exp((tmp)/dela)*pv(1,tmp+1); % INTGRL term
   end
%--------------------------------------------------------------------
   er(1, t+1) = sp(1, t+1) - cv(1, t+1);
   total_er = total_er + er(1, t+1);         % INTGRL term
end
h1=findobj('tag','cuc120');close(h1);
figure('tag','cuc120','Resize','on','MenuBar','none',...
   'Name','CUC120.m (Figure 1: PID control)',...
   'NumberTitle','off','Position',[160,80,520,420]);
ic = 0:Tot_hrs;
if pid == 1
   plot(ic/24, sp(1, :), ic/24, cv(1, :),'g+-', ic/24, er(1, :),'r*-');
   xlabel('Time elapsed, day');  axis([-inf inf -1 2]);
elseif pid > 1
   plot(ic, sp(1,:), ic, cv(1,:),'g+-', ic, er(1,:),'r*-');
   xlabel('Time elapsed, hr'); axis([-inf inf -0.3 1.5]);
   set(gca,'xtick',[0 6 12 18 24 30 36 42 48],...
```

```
                'ytick',[-0.2 0 0.2 0.4 0.6 0.8 1 1.2]);
end
tit=([logic ' control at kk=' num2str(kk) ', Tau=', num2str(tau)]);
title(tit);ylabel('Relative Temperature, ^oC');
grid on;    legend('Set Point','Response','Error');
```

Figure 9.4. Program to simulate PID control for floor-heating greenhouses **(CUC120.m)**.

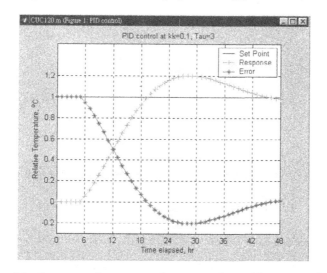

Figure 9.5a. Response and error curves for unit step change (showing overshoot).

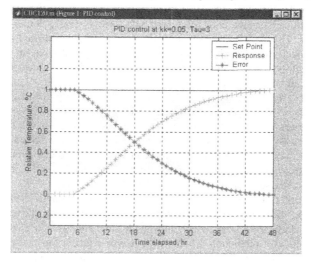

Figure 9.5b. Response and error curves for unit step change (showing undershoot).

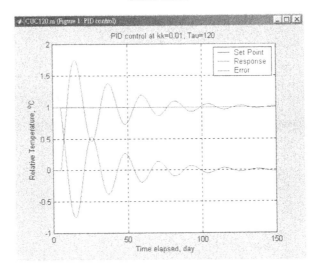

Figure 9.5c. Response and error curves for unit step change with long delay.

9.5. FEEDBACK VS. FEEDFORWARD CONTROL

9.5.1. Feedback and feedforward

Feedback control logic is well-established in both strategy and hardware configuration for many process controls. PID logic is one of the typical feedback control methods and is widely used. On the other hand, feedforward control is usually mentioned briefly in control textbooks, but is rarely found in practical applications. The main reason is that feedforward control depends on some kind of predictive model, even for physical systems, and works well alone in the practical sense. Therefore, the best solution is the combination of feedback and feedforward control if the latter is necessary.

When the system has a large time lag, as with floor heating, feedback techniques cannot effectively minimize the errors caused by the time lag in the system. This situation requires a form of predicted action. In feedforward control, the controller acquires information about potential upsets which have not yet affected the behavior of the process, anticipates the effect of these upsets on the process, and counteracts them before they are manifested on the process. The feedforward technique offers a potentially better solution for a system with large heat mass.

In the present section, it is easily shown that conventional feedback control applied to floor-heated greenhouses results in a large delay from the set point when floor heating starts and a large overshoot when heaters are turned off. Feedforward control introduces a prediction scheme, which will regulate the rate of heat input prior to the expected time and give complete control in this case because the system is perfectly predictable (see Takakura, *et al.*, 1993).

9.5.2. System response and its application to feedforward control

One well-known technique for predicting the dynamic response of a linear system to an arbitrary change of an input function is the weighing function method. In this method, the responses to all inputs are obtained by the principle of superposition. The theory of superimposition can be applied to any linear system.

Using Duhamel's integral, the response $R(t)$ to an arbitrary function $F(t)$ using the weighing function $W(\tau)$ is as follows:

$$R(t) = \int_0^t F(t - \tau) * W(\tau)\, d\tau \qquad (9.9)$$

where $W(\tau)$ is the derivative of a unit response.

If the unit response to the floor water temperature is defined as $Rf(t)$, the inside air temperature change $Ti(t)$ due to floor water is expressed as

$$Ti(t) = \int_0^t Q(t - \tau) * Rf'(\tau)\, d\tau \qquad (9.10)$$

where the superscript (') means the first derivative with respect to time. From the Laplace transform of the convolution integral, and by taking the Laplace transform of eq. 9.10, the following equation is given;

$$Q(s) = Ti(s) / Rf'(s) \qquad (9.11)$$

9.5.3. A floor-heating greenhouse with ideal conditions (CUC122)

The present model is very simplified and hypothetical, but adequate for showing the difference between feedback and feedforward logics. The unit step response of the inside air temperature due to the step change of floor water temperature is expressed by eq. 9.4 with a dead time and delay ($Rf(s)$ is equal to the Laplace transform of $RESF(t)$ in eq. 9.4). In the present model, heat transfer through the cover is assumed instantaneous and no response equation is needed for outside air temperature change. The delay of the control system is also introduced by the %---[DELAY] section as listed in Fig. 9.4. The system is well-defined and all responses are defined by analytical equations.

Now, the set point of inside air temperature $Ti(t)$ is set to

$$Ti(t) = K33 + A33 * \sin(OMEGA * t) \qquad (9.12)$$

After taking the Laplace transform of eqs. 9.4 and 9.12, the results are substituted into eq. 9.11. The inverse Laplace transform of eq. 9.11 gives finally the set point of floor water temperature in order to realize the inside air temperature change defined by eq. 9.12.

$$Q(t) = K33 + A33 * OMEGA * DELA * cos(OMEGA * (t + TAU))$$
$$+ A33 * sin(OMEGA * (t + TAU)) \qquad (9.13)$$

The program to compare the two control logics is shown in Fig. 9.6 and a typical result is indicated in Fig. 9.7. Feedback logic only follows the error which is detected at the present time. If the system delay is very large, as in the present case, the feedback system cannot catch up with the error in principle. A one or two hours delay is common in practical large-scale greenhouses. A larger gain in a feedback system can give a quicker response, but it brings larger amplitude difference in the response curve. In the present case, the system response curve is defined without any error, and gives a perfect fit with the set point for feedforward control after a certain time lapse. The %---[Delay] section of the source code listed in Fig. 9.6a provides comparison on feedback (PID) and feedforward control. The approach in deriving **CV** is the same as in **CUC120**, and the approach in deriving **CVF** is more accurate. The **quad8** or **quadl** function is used to do the integration from 0 to **t-TAU**. The function **quad8** was replaced by **quadl** in **MATLAB** from version 6 (Release 12).

```
% Control function for floor-heating greenhouse            CUC122.m
% Comparison of feedback (PID) and feedforward control
function cuc122
global TAU; global  DELA;
KK = 0.1;
TL = 0.9;   TD = 0.5;   TAU = 3.0; DELA = 5.0; % unit of time is hr
totnum=48; totnum2=totnum+1;
SP=zeros(1, totnum2); ER=zeros(1, totnum2);  PV=zeros(1, totnum2);
CV=zeros(1, totnum2); CVF=zeros(1, totnum2);
TOTAL_RESP=0;
SP(1,1)=sin(2*pi/24*0);            SP(1,2)=sin(2*pi/24*1);
ER(1, 1) = SP(1, 1) - CV(1, 1);   ER(1, 2) = SP(1, 2) - CV(1, 2);
TOTAL_ER = ER(1,1) + ER(1, 2);
PV(1,2)=KK*(ER(1,1)+TD*(ER(1,1)-0)+1/TL*TOTAL_ER);
for T = 2: 1: totnum;
    CLOCK=mod(T,24);    SP(1,T+1)=sin(2*pi/24*CLOCK);
    PV(1,T+1)=KK*(ER(1,T)+TD*(ER(1,T)-ER(1,T-1))+1/TL*TOTAL_ER);
    % TOTAL_ER is the Integral term of ER
    % PV    = KK*(ER + TD*DERIV(ICD,ER) + 1.0/TL*INTGRL(ICI,ER))
    % Feedback (PID) control logic
%----[Delay]------------------------------------------------------
    if T<TAU
        CV(1, T+1) = 0;                  CVF(1, T+1) = 0;
    else
        TMP = T-TAU;
        CV(1, T+1) = exp(-TMP/DELA) / DELA * TOTAL_RESP;
        TOTAL_RESP = TOTAL_RESP + exp((TMP)/DELA)*PV(1, TMP+1);
    % CVF(1,T+1)=exp(-TMP/DELA) / DELA * quad8('feedf',0,TMP);
    % quad8 is obsolete at MATLAB version 6 (R12)
        CVF(1,T+1)=exp(-TMP/DELA) / DELA * quadl('feedf',0,TMP);
```

```
% quadl is not available prior to MATLAB version 6 (R12)
  End
%-------------------------------------------------------------------
  ER(1, T+1) = SP(1, T+1) - CV(1, T+1);
% Error (ER) is the diff. between set point (SP) and controlled value (CV)
  TOTAL_ER = TOTAL_ER + ER(1, T+1);
end
h1=findobj('tag','cuc122'); close(h1);
figure('tag','cuc122','Resize','on','MenuBar','none', 'Name',...
  'CUC122.m (Figure 1: Feedback control vs. Feedforward control)',...
  'NumberTitle','off','Position',[160,80,520,420]);
ic = 0:totnum;
plot(ic, SP(1,:),ic, CV(1,:),'k+-',ic, CVF(1,:),'r*-');
set(gca,'xtick',[0 6 12 18 24 30 36 42 48],'ytick',[-1 -0.5 0 0.5 1]);
axis([-1 totnum2 -inf inf]);
xlabel('Time elapsed, hr');        ylabel('Temperature, ^oC');
legend('Set Point','Feedback','Feedforward'); grid on;
```

Figure 9.6a. A model to demonstrate feedback and feedforward control logics for a floor-heating greenhouse (**CUC122.m**).

```
% Intgrl term of feedforward control                      feedf.m
function y=FEEDF(t)
% This part is a copy of the eqs. 8, 9, 10, and 11 in the paper by
% T. Takakura, et al., Trans. ASAE (Vol. 37, 939-945)
global TAU; global DELA;
  OMEGA=2*pi/24;    A33=1;    K33=0;
  TT=K33+A33*OMEGA*DELA*cos(OMEGA*(t+TAU))+A33*sin(OMEGA*(t+TAU));
  y = exp(t/DELA).*TT;
```

Figure 9.6b. Subprogram of CUC122 (**feedf.m**).

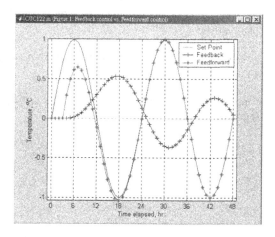

Figure 9.7. Air temperature changes controlled by feedback and feedforward logics with the set point temperature.

MATLAB FUNCTIONS USED

quad8	Numerically evaluate integral, higher order method.
	Q = QUAD8('F',A,B) approximates the integral of F(X) from A to B to within a relative error of 1e-3 using an adaptive recursive Newton Cotes 8 panel rule. 'F' is a string containing the name of the function. The function must return a vector of output values if given a vector of input values. Q = Inf is returned if an excessive recursion level is reached, indicating a possibly singular integral.
	Note: Function only available before and in version 5.3 (Release 11), obsolete after version 6.0 (Release 12).
quadl	Numerically evaluate integral, adaptive Lobatto quadrature.
	Q = QUADL(FUN,A,B) tries to approximate the integral of function FUN from A to B to within an error of 1.e-6 using high order recursive adaptive quadrature. The function Y = FUN(X) should accept a vector argument X and return a vector result Y, the integrand evaluated at each element of X.
	Note: Function available after version 6.0 (Release 12)

PROBLEMS

1.   Verify that eq. 9.3 is the general solution of eq. 9.2.

2.   Examine how the system changes if there is only PI and P control logic, modifying the program **CUC120**.

3.   Evaluate the effect of **KK** on the system response by re-running the program **CUC120** for several **KK** values.

4.   Change the input (**SP**) in the model **CUC120** from the step change to a periodic change and plot the result.

5.   Change the gain (**KK**) as well as time constants (**TD, TL**) in order to get better fit with the set point and explain the effect of each factor on the accuracy.

CHAPTER 10

PLANT RESPONSE TO THE ENVIRONMENT

10.1. INTRODUCTION

As mentioned in the preceding chapter, it is very important to include a plant growth sub-model in the total model. However, plant growth, particularly under cover, is complex, and modelling it is not simple. This is one of the main reasons why most greenhouse models do not have or have only rather simple sub-models of plant growth. In the open field, the time courses of environmental conditions such as air temperature and carbon dioxide concentration are smoother than those in some greenhouses where the objective is to optimize the environment. Basic response curves, such as photosynthetic and respiratory responses to light and temperature conditions, have been studied for many years and are useful for modelling plant growth in open fields. In protected cultivation such as heated greenhouses with CO_2 enrichment facilities, some conditions can be changed drastically and positively.

Studies on dynamic modelling of plant response to the environment started some years ago (*e.g.*, Takakura and Jordan, 1970). In those models, the static response curves, which are used for open field conditions were applied. A so-called black-box technique to find optimum environmental conditions for certain plant responses, such as photosynthesis and respiration, has also been studied (*e.g.*, Takakura *et al.*, 1978). The series of studies by Takakura and Jordan (1975) showed that the main physiological responses, such as photosynthesis and respiration, change considerably according to the dynamic change of environmental conditions, and that static response curves cannot be applied when environmental conditions can change drastically. Later this technique was called the "speaking plant technique", which means that environment conditions are optimized on the basis of plant responses. However, it can be said that there are not yet reliable reports on dynamic responses of plants to the environment, and that response curves obtained under static conditions are used for environments under cover where changes occur smoothly.

10.2. PLANT PHOTOSYNTHESIS AND RESPIRATION

Plant stomata control the flows of water vapor and carbon dioxide to and from plant leaves. The mechanism that opens and closes stomata is an interesting topic in simulation as well as in plant physiology. It is known that physical environments such as light, temperature and carbon dioxide affect their movement. A comprehensive model of their movement in terms of resistance is shown in Fig. 10.1 (Takakura *et al.*, 1975).

The components involved in stomatal resistance are shown in the figure separately. Resistance in the cuticle is parallel to that in stomata, and other resistances, such as boundary layer resistance on the leaf and internal resistance, are also connected serially. There are some experimental data which show that the stomatal resistance (**RS**) changes like an electrical switch based on the carbon dioxide level in the stomata (**CO2ST**). Another functional relationship found experimentally is that internal resistance (**RI**) is dependent on leaf temperature. Respiration (**RESP**), which takes place in the stomata, is a function of leaf temperature (**TP**) and is a well-known **Q10** function, which means that respiration doubles when temperature increases by 10°C. These functional relationships are expressed as:

$$RS = AFGEN\_RSTB(CO2ST) \qquad (10.1)$$

$$RI = AFGEN\_RITB(TP) \qquad (10.2)$$

$$RESP = Q10(TP) \qquad (10.3)$$

These expressions are different from mathematical ones, but are easily understood.

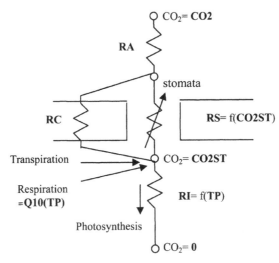

Figure 10.1. Basic structure of the model (after Takakura et al., 1975).

Another important relationship to be included is that of photosynthesis (**PHOTO**) and photosynthetically active radiation (**PAR**, W/m²). The relationship is expressed by two straight lines. If light is limiting the process, photosynthesis is expressed as

$$PHOTO = EFF * PAR \qquad (10.4)$$

On the other hand, if light is sufficient and carbon dioxide is limiting, then the equation becomes

$$\textbf{PHOTO} = 68.4 * \textbf{CO2ST} / \textbf{RI} \qquad (10.5)$$

where 68.4 is a unit conversion factor (from $\mu l/l$ CO_2 /m/s to kg CO_2 /ha/hr), **CO2ST** is CO_2 concentration in the stomata ($\mu l/l$), and **RI** is internal resistance (s/m). The unit of gas concentration is confusing. The unit 'ppm' is very popular but it does not indicate whether it is based on volume or weight. This is the reason why the unit 'vpm' is sometimes used to show clearly it is a volume basis. Then the effect of temperature on the volume and in the fixation of CO_2 must be taken into account; the concentration of CO_2 is often converted to photosynthesis products and expressed on a weight basis. In this sense, the unit 'μmol CO_2/mol air' is introduced. This is weight basis, but the value remains the same for mol basis at a fixed temperature. In the present textbook, as far as CO_2 is concerned, the effect of temperature on the conversion factors from volume to weight or vice versa is not taken into account.

Applying Ohm's law to the flow of CO_2 from the atmosphere, where its concentration is **CO2**, to inside the stomata, we have

$$\textbf{PHOTO - RESP} = 68.4 * (\textbf{CO2 - CO2ST}) / 1.6 / (\textbf{RA + RT}) \qquad (10.6)$$

where **RESP** is respiration (kg CO_2 /ha/hr), **RA** is boundary layer resistance, and **RT** is the total resistance of stomata and cuticle. The value 1.6 is introduced because of the difference in the diffusivities of water vapor and CO_2 gas. Stomatal resistance to CO_2 flow is 1.6 times larger than that to water vapor.

10.3. ENERGY BALANCE OF A PLANT LEAF

Energy balance equations for plant leaves are needed to find leaf temperature. First, it is assumed that the net radiation (**QNET**) on a leaf is the sum of sensible (**QS**) and latent heat (**QL**),

$$\textbf{QNET} = \textbf{QS} + \textbf{QL} \qquad (10.7)$$

Then, we need an approximation to relate the net radiation (**QNET**) and the photosynthetically active radiation (**PAR**), because **PAR** is the only variable to be defined as an energy source.

Let's assume **PAR** is half of the total solar radiation and make a radiation balance for plant leaves (also see section 5.2):

$$\textbf{QNET} = 2 * \textbf{PAR} * \textbf{ALP} + \textbf{QUL} * \textbf{EPSP} - 2 * \textbf{EPSP} * \textbf{SIG} * (\textbf{TP}+273)^4 \qquad (10.8)$$

where **ALP** is the absorptivity of the plant for solar radiation, **QUL** is the sum of long wave radiation from the atmosphere and the ground, and it is assumed that **PAR** is constant. **EPSP** is emissivity and absorptivity of plant leaves for long wave radiation and **SIG** is the Stefan-Boltzmann constant.

Sensible heat itself is expressed as

$$QS = CA * (TP - TA) / RA \qquad (10.9)$$

where **CA** is volumetric heat capacity of the air ($J/m^3/^\circ C$), **TP** is leaf temperature ($^\circ C$), and **TA** is ambient temperature. Then, it is clear that the heat transfer coefficient due to convection is **CA/RA**. On the other hand, the latent heat is expressed by using a very common expression in meteorology, that is, the psychrometric constant (**PSCH**, mb/$^\circ C$) as

$$QL = (PLS - PA) / (RA + RT) * CA / PSCH \qquad (10.10)$$

Again, the saturated water vapor pressure at the leaf temperature is obtained from the saturation curve shown in Fig. 4.9. In this case, another simple approximated exponential curve is used and it is in a function form:

$$PLS = PSYCR(TP) \qquad (10.11)$$

Eliminating **QNET, QS, QL**, and **PLS** from eqs. 10.7 to 10.10, and taking one **TP** term, which is first order, to the left-hand side of the equation, we have

$$TP = TA + (2 * PAR * ALP + QUL*EPSP - 2 * EPSP * SIG * (TP+273)^4 - CA * (PSYCR(TP) - PA) / PSCH / (RA + RT)) * RA / CA \qquad (10.12)$$

This is an implicit expression in terms of **TP** because the fourth order term and the function are still on the right hand side of the equation.

Therefore, the unknown variables in eqs. 10.1 through 10.6 and eq. 10.12 are **PHOTO, CO2ST, TP, RS, RI,** and **RESP**. Then, the total number of equations is six (eqs.10.4 and 10.5 are counted as one), although eq. 10.12 is an implicit expression.

10.4. STOMATAL RESISTANCE OF PLANTS (CUC151)

Instead of **IMPL** function, available in **CSMP**, **MATLAB** provides function **fzero** to solve the implicit form of equations. This part of the program is as follows:

```
%CUC151.m
..........
y=fzero('findco2ver1',CO2ST);
..........

%findco2ver1.m
```

```
function y=findco2ver1(CO2ST)
...
RS = AFGEN_RSTB(CO2ST);
RT = RC*RS/(RC+RS);
TP = TA+(2*PAR*ALP+QUL*EPSP-2*EPSP*SIG*(TP+273)^4 ...
     - CA*(PSYCR(TP) - PA)/PSCH/(RA+RT))*RA/CA;
RI = AFGEN_RITB(TP);
PHOTO = MIN (EFF*PAR, 68.4*CO2ST/RI);
Q10 = 2.0^((LIMIT(0.,40.,TP) - TREF)*0.1);
RESP = 5.0*Q10;
CEND= CO2 - (PHOTO - RESP)*1.6*(RT+RA)/68.4;
y=CO2ST - CEND;
```

Fzero is a scalar non-linear zero finding function. Y = **fzero**(FUN,Y0) tries to find a zero of FUN near Y0. FUN is usually an M-file. The value Y returned by **fzero** is near a point where FUN changes sign, or NaN if the search fails. **CEND** is the expression for **CO2ST** in an implicit form. If the difference of **CO2ST** and **CEND** is within an acceptable range, the implicit equation is solved. A more detailed description of the **fzero** function can be found at the end of this chapter.

The stomatal resistance **RS** is considered to change from its minimum value (**RSMN**) to its maximum value (**RSMX**) according to the aperture of the stomata; the aperture is controlled by the CO_2 concentration in the stomata (**CO2ST**), as shown in Fig. 10.2, and this is involved in the function **AFGEN_RSTB**. Internal resistance (**RI**) is also expressed by using **AFGEN_RITB**. The relationship between **RI** and **TP** can be approximated using a downward convex parabola. This indicated that **RI** has a minimum value at a certain temperature.

MIN is a function to find the minimum value among options listed in the parentheses and in this case **PHOTO** is set equal to the lower value of either **EFF*PAR** or 68.4*CO2ST/RI. This approximation is called Blackman's expression (see Fig. 10.6). Another approximation can be made using hyperbolic curves.

Leaf resistance consists of the resistances of stomata and cuticles in parallel; therefore the total resistance (**RT**) is as shown in the program.

The function **LIMIT** is used to limit the third argument -- **TP** in this case. If **TP** is outside the range between 0 and 40, then **TP** is set to the nearer boundary.

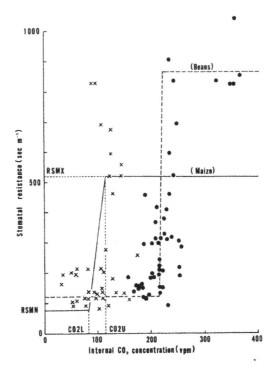

Figure 10. 2. Simplified relationship between stomatal resistance (s/m) and internal CO2 concentration (vpm) with experimental data (after Takakura et al., 1975).

```
%     Behavior model of stomata opening                          CUC151.m
%     Changes of RS and PHOTO by CO2
%     Enter 'cuc151' or 'cuc151(1)' in Command Window for Figs. 1 and 2,
%     enter 'cuc151(2)' for Figs. 3, 4 and 5.
%     Subprogram: findco2ver1.m
%
function cuc151(action)
if nargin==0,action=1;end;
switch action
case 1
  global rt photo co2 ra par
  global rs resp tp
  par1=[50 100 500];        %par: light level, in W/m2
  co2mat=0:20:1000;
  clc
  for k=1:length(par1)
  par=par1(k);   co2st=300;
  for j=1:51
      co2=co2mat(j);
%--------------------------------------------------------------------
      y=fzero('findco2ver1',co2st);  % core of this program
%--------------------------------------------------------------------
      clc
```

```
        fprintf('\nPAR=%3.0f CO2=%5.0f\n ',par,co2);
        rt1(k,j)=rt;             photo1(k,j)=photo;
        end
    end
    cuc151_fig1=findobj('tag','fig1'); close(cuc151_fig1);
    cuc151_fig2=findobj('tag','fig2'); close(cuc151_fig2);
    figure('tag','fig1','name','CUC151: 1.Photosynthesis','menubar', ...
        'none','NumberTitle','off','Position',[100,150,520,380]);
    cuc151_fig1=plot(co2mat,photo1(1,:) ,'r*-',co2mat,photo1(2,:),...
        'b^-',co2mat,photo1(3,:) ,'ko-');
    set(cuc151_fig1,'linewidth',1);
    axis([0 1000 0 100]);
    xlabel('CO_2 in the air (ul/l)');
    ylabel('Photosynthesis (kgCO_2/ha/hr)');
    legend('PAR=50W/m^2','PAR=100','PAR=500',2);
    grid on;
    figure('tag','fig2','name','CUC151: 2.Stomatal Resistance',...
      'menubar','none','NumberTitle','off','Position',[140,120,520,380]);
    cuc151_fig2=plot(co2mat,rt1(1,:) ,'r*-',co2mat,rt1(2,:),...
            'b^-',co2mat,rt1(3,:) ,'ko-');
    set (cuc151_fig2,'linewidth',1);
    axis([0 1000 0 600]);
    xlabel('CO_2 in the air (ul/l)');
    ylabel('Total Resistance of Stomata & Cuticle (s/m)');
    legend('PAR=50W/m^2','PAR=100','PAR=500',4);
    grid on;
case 2
    global rt photo co2 ra par
    global rs resp tp
    co2=350;
    par1=0:10:500;
    clc
    for k=1:length(par1)
        par=par1(k);   co2st=300;
  %-----------------------------------------------------------------
        y=fzero('findco2ver1',co2st);
        % find root for eqs. listed in findco2ver1.m
  %-----------------------------------------------------------------
        clc
        fprintf('\nPAR=%3.0f PHOTO=%3.0f RS=%4.1f TP=%3.1f  resp=%4.2f\n'...
                    ,par,photo,rs,tp,resp);
        rs1(k)=rs;            rt1(k)=rt;
        photo1(k)=photo;   resp1(k)=resp;               tp1(k)=tp;
        end
    cuc151_fig3=findobj('tag','fig3');close(cuc151_fig3);
    figure('tag','fig3','name','CUC151: 3.Given Constant CO2', 'menubar', ...
        'none','NumberTitle','off','Position',[100,150,520,380]);
    plot(par1,resp1,par1,tp1)
    axis([0 500 0 100]);
    hold on;
    plotyy(par1,photo1,par1,rs1);
    tit=(['Given Tair=30 degree C & CO2=' num2str(co2)]);
    title(tit);
    xlabel('PAR (W/m^2)');
    ylabel('Photosynthesis (kgCO_2/ha/hr), RESP, TP(^oC)');
    text(200,60,'Photosynthesis');
    text(10,85,'Stomatal resistance');
    text(200,27,'Leaf temperature');
    text(200,32,'Air temperature=30');
    text(300,12,'Respiration');
    text(540, 15, 'Stomatal Resistance,(RS) in s/m', 'Rotation', 90);
```

```
    line([0 500],[30 30],'color',[1 0 0],'linestyle',':');
    hold off
    cuc151_fig4=findobj('tag','fig4');close(cuc151_fig4);
    figure('tag','fig4','name','CUC151: 4. Net Photosynthesis', …
      'menubar','none','NumberTitle','off','Position',[120,130,520,380]);
    plot(par1,photo1,'r*-',par1,resp1,'b^-',par1,photo1-resp1,'ko-')
    xlabel('PAR (W/m^2)');
    ylabel('CO_2 released or absorbed (kgCO_2/ha/hr)');
    legend('Photosynthesis','Respiration','Net Photosynthesis',4);
    %
    cuc151_fig5=findobj('tag','fig5');close(cuc151_fig5);
    figure('tag','fig5','name','CUC151: 5. RS and RT','menubar', …
          'none','NumberTitle','off','Position',[120,130,520,380]);
    plot(par1,rs1,'r*-',par1,rt1,'k^-','linewidth',2);
    xlabel('PAR (W/m^2)');      ylabel('Resistance, (s/m)');
    title('Given RC is 4000 s/m');
    legend('Stomatal resistance','Total resistance (Stomata + Cuticle)');
End
```

Figure 10.3a. Main program to simulate stomatal resistance **(CUC151)**.

```
%                                                          findco2ver1.m
function y=findco2ver1(co2st)
global rt photo co2 ra par
global rs resp tp
   aa=0; bb=0;
   eff=0.62;      % eff: slope of the photosynthesis curve at PAR=0,
                  % leaf light utilization efficiency, in umol CO2/umole photon
   tref=20;       % reference temperature, in degree C
   ta=30;         % Air temperature, in degree C
   alp=0.8;       % Absorptivity of plant leaf for solar radiation
   epsp=0.95;     % Emissivity/Absorptivity of plant leaf for long wave  rad.
   sig=5.67;      % Stefen-Boltzman constant
   rh=0.75;       % Relative Humidity is 75%
   ta=30;         % ta: air temperature (degree C)
   qul=150;       % qul: longwave rad. for the atmosphere and the ground (W/m2)
   ra=10;         % ra: diffusion resistance in laminar layer (s/m)
   rc=4000;       % rc: cuticle resistance
   ca=1164;       % ca: volumetric heat capacity of air at constant pressure
   psch=0.67;     % psch: psychrometric constant (mb/C)
   tp=ta;
   %---------------- co2st=IMPL(300,0.01,cend)-----------------------
   rs=afgen_rstb(co2st);
   rt=rc*rs/(rc+rs);      % total resistance for water vapor of leaves (s/m)
   tp4=((273.16+tp)/100)^4;
   aa=(2*par*alp+qul*epsp-2*epsp*sig*(tp4))*ra/ca;
   bb=ca*((psycr(tp)-rh*psycr(ta))/psch/(ra+rt))*ra/ca;
   tp=ta+aa-bb;               ri=afgen_ritb(tp);
   photo=min(eff*par,68.4*co2st/ri);
   tp=limit(0,40,tp);       Q10=2^((tp-tref)*0.1);      resp=5*Q10;
   cend=co2-(photo-resp)*1.6*(rt+ra)/68.4;
   y=co2st-cend;
%--------------------------------------------------------------------
function vapres_value=psycr(temp)
    vapres_value=6.11*exp(17.4*temp/(239+temp));
% Saturated vapor pressure as a function of air temperature, in mb
%--------------------------------------------------------------------
function rs_value=afgen_rstb(co2st)
%  rstb_x=[0 80 120 1000];
%  rstb_y=[74.4 74.4 521 521];
```

```
%    rs=interp1(rstb_x,rstb_y,co2st,'linear');
% Another approach
if co2st<=80,
   rs_value=74.4;
elseif co2st>=120,
   rs_value=521;
else
    rs_value=74.4+(521-74.4)*(co2st-80)/(120-80);
end
%------------------------------------------------------------------
function tp_value=limit(min,max,tp)
if tp<min,
   tp_value=min;
elseif tp>max,
   tp_value=max;
else
   tp_value=tp;
end
%------------------------------------------------------------------
function ri_value=afgen_ritb(tp)
   ritb_x=[12 17 22 27 32 37 42]; % all in degree C
   ritb_y=[250 170 130 110 100 120 160];
   if tp<=12,
      ri_value=250;
   elseif tp>42,
      ri_value=160;
   else
       ri_value=interp1(ritb_x,ritb_y,tp,'*linear');
       % when intevals on ritb_x, are the same, *linear can be used
   end
```

Figure 10.3b. Subprogram to simulate stomatal resistance (**findco2verl.m**).

The main and subprograms are shown in Figs. 10.3a and 10.3b, respectively. In total, five functions were included in the subprogram (Fig. 10.3b) including **FINDCO2VERL, PSYCR, AFGEN_RSTB, LIMIT,** and **AFGEN_RITB.**

The function **PSYCR** is used to calculate saturated vapor pressure as a function of air temperature. A simpler and rougher expression than the one introduced in section 4.5 is used. In the present simulation the independent variable is **CO2.**

RS changes linearly according to **CO2ST** between its minimum (74.4) and maximum values (521) as shown in function **AFGEN_RSTB**. **RI** values can also be found in function **AFGEN_RITB**. In between minimum and maximum values, **RI** can be derived using function **interp1(X, Y, XI, method)**. All the interpolation methods require that the vector **X** be monotonic. **X** can be non-uniformly spaced. When **X** is equally spaced and monotonic, the methods '*linear', '*cubic' or '*nearest' can be used for faster interpolation.

Two figures as shown in Fig. 10.4a and 10.4b can be generated after entering 'cuc151' or 'cuc151(1)' in the Command Window. Fig. 10.4a shows the photosynthesis in relation to external CO_2 concentration. Fig. 10.4b shows the stomatal resistances change due to the changes of CO_2 concentration in the air. Three different curves are given for different light levels. It is not indicated in this figure, but a shift in the transition range of internal CO_2 concentration from lower to higher -- from 80 to 200 $\mu l/l$, for example -- causes the **RS** level to be steadily lower at a high light level.

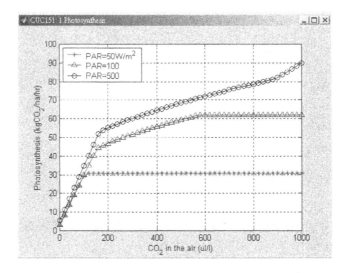

Figure 10.4a. Photosynthesis in relation to external CO_2 concentration: (First output figure after entering cuc151(1) in Command Window).

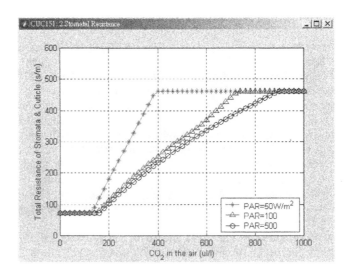

Figure 10.4b. Stomatal resistance in relation to external CO_2 concentration: (Second output figure after entering cuc151(1) in Command Window).

Three more figures can be generated after entering 'cuc151(2)' in the Command Window.   Fig. 10.5a shows photosynthesis, stomatal resistance, respiration and leaf temperature under various PAR.

Gaastra (1959) reported the stomatal response of turnips to light intensity.   The rapid change of stomatal resistance under lower light intensity and the shift of the occurrence of its drastic change to higher CO_2 range due to high light intensity are in good agreement with the simulated results (see Takakura et al., 1975).

The response curves of photosynthesis in Fig. 10.4b are not consistent according to the change of light levels.   When light level is low, the photosynthetic response is very similar to that of light change, that is, Blackman's curve.   When **PAR** is 100 W/m^2, the sharp edge from the CO_2 limiting range to the saturated range is smoothed compared to the curve for **PAR** of 50 W/m^2.   The shape of the curve is strange when **PAR** is 500 W/m^2.   Sharper increase in photosynthetic CO_2 uptake appears after the stomata are completely closed due to the upper limit of resistance.   This phenomenon has not been experimentally verified yet.   The experimental data are mostly scattered around these curves, but this kind of phenomenon cannot be substantiated if we do not know the theory.   This is one of the definite reasons why simulation is one of the two wheels of a cart; the other wheel is field experiments.

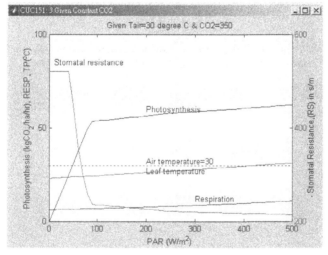

Figure 10.5a. Simulated results of photosynthesis, stomatal resistance, respiration and leaf temperature under various PAR: (First figure after entering cuc151(2)) (after Jordan, 1992).

In Fig. 10.5a, it is clearly shown that the original shape of Blackman's curve for photosynthesis is modified to a hyperbolic type to some extent due to the small change of stomatal resistance.   When light level is low, stomata are closed.   It is often said that photosynthesis is light-dependent when light is the limiting, but it can be explained by the fact that stomata are closed at low light level.   Actually stomata are the limiting factor.   Leaf temperature is lower than the air temperature but leaf temperature monotonously increases with light increase.   The same tendency occurs in leaf respiration.

Fig. 10.5b shows photosynthesis, net photo- synthesis and respiration, and Fig.10.5c shows stomatal resistance and total resistance (stomatal and cuticle).

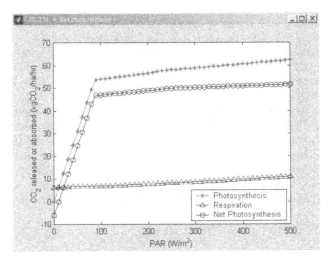

Figure 10.5b. Simulated results of (net) photosynthesis and respiration under various PAR: (Second figure after entering cuc151(2) in Command Window).

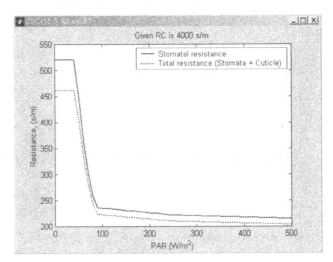

Figure 10.5c. Simulated results of stomatal resistance due to PAR change: (Third output figure after entering cuc151(2) in Command Window).

10.5. PLANT GROWTH MODEL

10.5.1. General concept of plant growth and yield models

Plant growth is an example of a set of non-linear dynamic systems and has been studied for many years. Plant vegetative growth, development, and production phases have been studied by the Dutch group led by de Wit (*e.g.*, 1972; 1974). Continuous research efforts have been conducted in this area, and the same methodology has been expanded from open-field crops to greenhouses (Jones *et al.*, 1991; Jones, 1991). In general, plant growth which is continuously changing can be expressed as systems of differential equations such as the system of equations in section 3.3.2.

The transition from vegetative to reproductive stages in crops can also be modelled in a similar way. For example, introducing the rate of development (**DVS**), which is assumed to be a function of temperature (**TEMP**) and daylength (**DAYL**),

$$d\textbf{DVS(t)} / dt = f(\textbf{TEMP, DAYL}) \tag{10.13}$$

If **DVS** is assumed to be a function of temperature alone, the following expression could be used (Horie, 1987):

$$d\textbf{DVS(t)}/dt = 1.0 - exp(-\textbf{KD*(TEMP - TCD})) \quad \text{when } \textbf{TEMP} \geq \textbf{TCD} \tag{10.14}$$

$$= 0 \quad \text{when } \textbf{TEMP} < \textbf{TCD} \tag{10.15}$$

where **TEMP** is daily temperature, **TCD** is a minimum temperature for development, and **KD** is a parameter. Then, **DVS** is 1.0 at flowering and is between 0 and 1 during the vegetative phase.

10.5.2. Dry matter production model (CUC160)

Plant models, in general, cover a long period of time from seedling to harvest, so the time period for most plant models is several months. Each reaction is described by a differential equation. The main consideration is to attain agreement between the model and experimental data over the entire period of time. The main purpose of the present plant modelling is not to see hourly changes, although hourly changes can be seen through the same model. A simple model offered by Jones (1991) is considered.

The basic equation in Jones' model describes the change in dry matter production which is related to photosynthesis and respiration by the expression

$$d\textbf{WGT} / dt = \textbf{E * (PHOTO} - \textbf{RESP * WGT}) \tag{10.16}$$

where $d\mathbf{WGT}/dt$ is the rate of dry matter production of the plant (g tissue/m^2/hr), **WGT** is total plant dry weight (g/m^2), **E** is conversion efficiency of CH$_2$O to plant tissue (g tissue/g CH$_2$O), and **PHOTO** is canopy gross photosynthesis rate (g CH$_2$O/m^2/hr).

Gross photosynthesis is expressed by the well-known formula

$$\mathbf{PHOTO} = \frac{\mathbf{PMAX*EFF*PAR}}{\mathbf{PMAX+EFF*PAR}} \qquad (10.17)$$

where **PMAX** is the maximum rate of photosynthesis to be attained and is a function of CO$_2$ concentration and leaf temperature, **EFF** is the slope of the photosynthesis curve where **PAR** is zero, and **PAR** is the photosynthetically active radiation on a leaf. Thus, a typical expression for **PMAX** is the product of **PHI*CO2** and **PTM** where **PHI*CO2** is the amount of CO$_2$ flow, **PTM** is a function of leaf temperature, and **PHI** is leaf conductance to CO$_2$ transfer. Eq. 10.17 is another form of Blackman's expression, which was shown in the preceding section. A general curves resulting from these equations and a **MATLAB** program are shown in Figs. 10.6a and 10.6b.

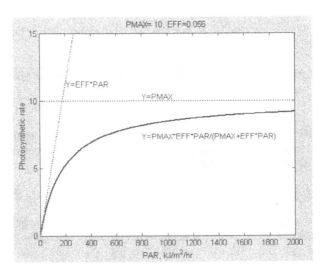

Figure 10.6a. Program output showing Blackman's expression (**CUC155.m**).

```
% Blackman's Expression                                          CUC155.m
par0=2000;  %  (kJ/m2/hr)
EFF=0.056;  % slope
PMAX=10;      % Max. photosynthetic rate at CO2 and Light saturation points
t=1:1:par0;
for par=1:par0;
    photo(par)=PMAX*EFF*par/(PMAX+EFF*par);     photo2(par)=EFF*par;
```

```
end
h1=findobj('tag','Blackman');close(h1);
figure('tag','Blackman','Resize','on','MenuBar','none',...
   'Name','CUC155: Blackman''s Expression',...
   'NumberTitle','off','Position',[160,80,520,420]);
plot(t,photo,'k-',t,photo2,'r:','linewidth',2);
line([0 par0],[PMAX PMAX],'linestyle',':','linewidth',2);
axis([-inf inf 0 PMAX*1.5]);
titleline=['PMAX= ' num2str(PMAX) ', EFF=' num2str(EFF)];
title (titleline);
xlabel('PAR, kJ/m^2/hr');        ylabel('Photosynthetic rate');
text(800,PMAX*1.03,'Y=PMAX'); text(205,EFF*200,'Y=EFF*PAR');
text(800,0.9*PMAX*EFF*800/(PMAX+EFF*800),'Y=PMAX*EFF*PAR/(PMAX+EFF*PAR)');
```

*Figure 10 6b. Program to illustrate Blackman's expression (**CUC155.m**).*

Light penetration in the plant canopy has been analyzed for many years. One of the typical representations for downward light at a certain level in a plant canopy is derived from Lambert-Beer's law as,

$$PAR = PAR0 * f(L) = PAR0 * \exp(-K*L) \qquad (10.18)$$

where **PAR0** is the **PAR** at the top of the canopy, **K** is the extinction coefficient of the canopy, and **L** is the leaf area index from the top to the chosen level in the canopy. The absorbed fraction of light, d**PAR**, is the product of leaf fraction, d**L**, and **PAR** which is not transmitted. **LT** is the light transmission coefficient of leaves. The equation for d**PAR** is

$$-d\mathbf{PAR} = d\mathbf{L} * (1 - \mathbf{LT}) * \mathbf{PAR} \qquad (10.19)$$

Then,

$$\mathbf{PAR} = -\frac{1}{1-\mathbf{LT}} * \frac{d\mathbf{PAR}}{d\mathbf{L}} = -\frac{\mathbf{PAR0}}{1-\mathbf{LT}} * \frac{df(\mathbf{L})}{d\mathbf{L}}$$

$$= \frac{\mathbf{PAR0}}{1-\mathbf{LT}} * K * \exp(-K*L) \qquad (10.20)$$

Combining eqs. 10.17 and 10.20, and integrating over the entire leaf area of the canopy gives the rate of canopy net photosynthesis per unit ground area as

$$\begin{aligned}\mathbf{PHOTO} = D\,/\,K * PHI * CO2 * PTM * \log((EFF * K * \\ PAR0 + (1 - LT) * PHI * CO2 * PTM)\,/\,(EFF * K \\ * PAR0 * \exp(-K * LAI) + (1 - LT) * PHI * CO2 * \\ PTM)) \qquad (10.21)\end{aligned}$$

where **D** is the coefficient to convert photosynthesis calculations from the unit μmol $CO_2/m^2/s$ to the unit g $CH_2O/m^2/hr$, **PHI** is leaf conductance to CO_2 flow (mol air/m^2/s), **CO2** is CO_2 concentration of the air (μmol CO_2/mol air), **PTM** is a dimensionless function of temperature, **EFF** is leaf light utilization efficiency (μmol CO_2/μmol photon), **K** is the canopy light extinction coefficient, **PAR0** is light flux density at the top of the canopy (μmol photon/m^2/s), and **LAI** is the canopy leaf area index (m^2 leaf/m^2 ground).

The function **PTM** expresses the effect of leaf temperature (**TP**) on the maximum rate of photosynthesis for a single leaf,

$$\textbf{PTM} = 1 - ((\textbf{TPH} - \textbf{TP}) / (\textbf{TPH} - \textbf{TPL}))^2 \qquad (10.22)$$

where **TPH** is the temperature at which leaf photosynthesis is maximum, and **TPL** is the temperature below which leaf photosynthesis is zero.

The leaf area index is

$$\textbf{LAI} = \textbf{NP} * \textbf{DELTA} / \textbf{BETA} * \log(1 + \exp(\textbf{BETA} * (\textbf{NL} - \textbf{NB}))) \quad (10.23)$$

where **LAI** is leaf area index (m^2 leaf/m^2 ground), **NP** is plant density (number of plants/m^2), **NL** is leaf number, and **DELTA**, **BETA** and **NB** are empirical coefficients.

Maintenance respiration is a function of leaf temperature and is expressed as

$$\textbf{RESP} = \textbf{RESP25} * \exp(0.0693*(\textbf{TP} - 25)) \qquad (10.24)$$

where **RESP** is the maintenance respiration rate (g CH_2O/g tissue/hr), and **RESP25** is the respiration rate at 25 °C (g CH_2O/g tissue/hr).

The state variable for number of leaves on a vegetable plant (**NL**) is assumed to be a function of temperature:

$$d\textbf{NL} / d\text{t} = \textbf{RESP} * \textbf{AFGEN\_RTP(TP)} \qquad (10.25)$$

where **AFGEN_RTP(TP)** is a function of leaf temperature (**TP**) as shown in Fig. 10.7.

Two differential equations, eqs. 10.16 and 10.25, and four functional relationships (eqs. 10.21 - 10.24) connect the unknown variables. There are six unknown variables, **WGT**, **PHOTO**, **RESP**, **PTM**, **LAI**, and **NL**, and three inputs (boundary conditions), **PAR0**, **CO2**, and **TP**. All other terms are parameters or constants that should be specified by the programmer. The complete listing of the model is in Fig. 10.8, and the result of a two-day run is shown in Fig. 10.9.

This model can easily be combined with any of the greenhouse models such as those described in Chapter 6.

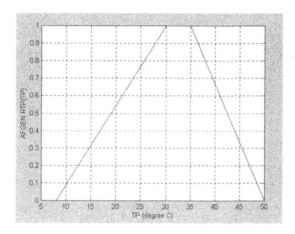

Figure 10.7. Effect of temperature on number of leaves (after Jones, 1991).

```
%    Plant Growth Model                                      CUC160.m
%    also requires function: soil160.m
%
clear all;clc
t0=0;tfinal=48;
y0=[0; 0];           % initial condition of WGT and NL
[t,y]=ode15s('soil160',[t0 tfinal],y0);
h1=findobj('tag','Plant Growth'); close(h1);
figure('tag','Plant Growth','Resize','on','MenuBar','none',...
   'Name','CUC160: Example of plant dry weight increase',...
   'NumberTitle','off','Position',[160,80,520,420]);
subplot(1,2,1); plot(t,y(:,1)*1000);
xlabel('Time (hr)');   ylabel('Dry matter Production (mg tissue/m^2/hr)');
axis([0, 50, 0, 54]);
set(gca,'ytick',[18 36 54]);  grid on;
subplot(1,2,2); plot(t,y(:,2));
xlabel('Time (hr)');   ylabel('Leaf Number');
axis([0, 50, 0, 0.005]);
set(gca,'ytick',[0.001 0.002 0.003 0.004 0.005]);    grid on;
fprintf('\n\n');
disp('Thank you for using'); disp(' ');
disp('CUC160: a plant growth model.');   disp(' ');
```

Figure 10.8a. Main program of plant dry weight increase model **(CUC160)**.

```
%    Subprogram for model CUC160                               soil160.m
%    Also requires functions TABS.m and SOLAR.m  (refer to Chapter 4)
function dy = soil160(t,y)
RP=500;                       % RP: Solar radiation amp (kJ/m2/hr)
OMEGA=2.0*pi/24.0;            % Time (hr)
clk=mod(t,24);               % time (24 hr basis)
T0=10.0; TU=5.0; TBL=10.0;   % Temp (C)
TP = T0 + TU*sin(OMEGA*(clk-8));  % Temp (C)
PAR0= SOLAR(RP,OMEGA,clk);    % calling function SOLAR()
% PAR0: Light flux density at the top of the canopy (umol photon/m2/s)
```

```
CO2=350+50*sin(OMEGA*(clk+6));
% CO2: CO2 concentration of the air (umol CO2/mol air, ppm, ul/l)
E=0.7; % conversion efficiency of CH2O to plant tissue (g tissue/g CH2O)
% WGT: total plant dry weight (g/m2)
% PHOTO: canopy gross photosynthesis rate (g CH2O/m2/hr)
D=0.108;          % coefficient to convert photosynthesis calculations
                  % from umol CO2/m2/s to g CH2O/m2/hr
K=0.58;           % canopy light extinction coefficient
PHI=0.0664;       % leaf conductance to CO2 (umol CO2/umol air m2/sec)
EFF=0.056;        % leaf light utilization efficiency (umol CO2/umol photon)
LT=0.002;         % light transmission coefficient of leaves
NP=4;             % plant density, (Number of plants/m2)
DELTA=0.074;BETA=0.38;NB=13.3;    % empirical coefficient
TPH=30;           % Temperature at which leaf photosynthesis is maximum
TPL=5;            % Temperature below which leaf photosynthesis is zero
PTM=1-((TPH-TP)/(TPH-TPL))^2; % dimensionless function of temperature
WGT=y(1);         % y(1) is dry weight increase (WGT)
NL=y(2);          % y(2) is number of leaves (NL)
LAI=NP*DELTA/BETA*log(1+exp(BETA*(NL-NB)));
% LAI: canopy leaf area index (m2 leaf/m2 ground)
PHOTO=D/K*PHI*CO2*PTM*log((EFF*K*PAR0+(1-LT)*PHI*CO2...
   *PTM)/(EFF*K*PAR0*exp(-K*LAI)+(1-LT)*PHI*CO2*PTM));
% Growth and maintenance respiration
RESP25=0.0006; % respiration rate at 25 degree C (g CH2O/g tissue/hr)
RESP=RESP25*exp(0.0693*(TP-25));
% RESP: maintenance respiration rate (g CH2O/g tissue/hr)
WGT=E*(PHOTO-RESP*WGT);          % dry weight increase
RTP= AFGEN_RTP(TP);              % function AFGEN_RTP
NL=RESP*RTP;                     % NL: leaf number increase
dy=[WGT; NL];
%--------------------------------------------------------------
function RTP= AFGEN_RTP (TP)
%TP= [8  12  30  35  50];
%RTP=[0 0.55  1  1   0];
if TP<=8 | TP>=50,RTP=0;
elseif TP<=12, RTP=(TP-8)/(12-8)*0.55;
elseif TP<=30,RTP=0.55+(TP-12)/(30-12)*(1-0.55);
elseif TP<=35,RTP=1;
elseif TP<50,RTP=1-(TP-35)/(50-35)*1;
end
```

Figure 10.8b. Subprogram of plant dry weight increase model (**Soil160**).

It is apparent that the plant weight increase is continuous, but the present model is too rough to express its hourly change. Leaf number is also expressed as a continuous function, but it is in reality discrete. Therefore, the outputs shown in Fig 10.9 are hypothetical. However, for a longer range of time, that is, for the whole growth period, it would give reasonable results. If daily changes of weather can be supplied, longer term of prediction can be attained.

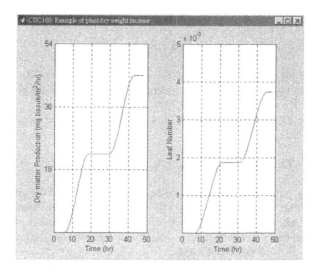

Figure 10.9. Simulated results of dry weight increase and number of leaves produced over time (Output figure after entering cuc160 in command window).

MATLAB FUNCTIONS USED

Fzero   If f(x) is a function of a single variable and is coded in an M-file function with the name **fname**, and if **k0** is a guess for a root of this function, then **fzero** will attempt to find the actual root of the function to a tolerance, **tol**.   The implementation is **kroot = fzero('fname', k0, tol)**.   If the third item in the argument list, **tol**, is omitted, a tolerance equal to **eps**, the machine accuracy, is used.   If there is a nonzero fourth item in the list, intermediate steps in the calculation will be printed.   The algorithm searches for a change in sign of the function and uses quadratic interpolation near the root.

PROBLEMS

1.   Derive the conversion factor 0.54 from the units $\mu m/s$ m/s to the units $kgCO_2/ha/hr$.

2.   At a fixed temperature, verify that CO_2 concentration in the air of 350 ppm is equal to 350 μmol CO_2/ mol air.

3.   Derive eq. 10.12 from eqs. 10.7- 10.11.

4.   Show the relationship between **KM** in eq. 4.14 and **PSCH** in eq. 10.10.

5.   Explain why **RS** cannot be solved explicitly.

6.   Change the transient region of CO_2 from $80 - 120$ $\mu l/l$ to $200 - 240$ $\mu l/l$ in function **AFGEN_RSTB** in the program **Soil151** and rerun the program.

7.   Change the programs **CUC151** and **Soil151** in order to obtain the result shown in Fig. 10.6.

8.   Use **MATLAB** to solve the following equation: $x = \tan(x) - 1.0$

9.   Write a program to calculate wet-bulb temperature (**TOW**) given dry-bulb temperature (**TO**) and humidity ratio (**WO**) by using the **fzero** function of **MATLAB**.   Then, from dry-bulb and wet-bulb temperatures, find humidity ratio and dew point temperature.   Note: Use the following equations:
WWS   =   **FWS(TOW)**, and
TOW   =   **TO - (2501.0 - 2.38*TOW)*WWS + (2501.0 + 1.805*TO - 4.186*TOW)*WO**
where **FWS** is a function to calculate the saturated humidity ratio in Fig. 4.9c.

10.   Derive eq. 10.21, using the formula $\displaystyle\int_0^x \frac{f'(x)}{f(x)}\,dx = \log f(x)$

11.   Plot **PHOTO** by using eq. 10.17 as a function of **PAR** with **CO2** as a parameter and the remaining terms as constants.

12.   Explain why the unit of **PHI** is mol air/m^2/s.

13.   The equation **PMAX = PHI*(CO2 - 0.0)** shows that the maximum CO_2 flow is the product of CO_2 flow conductance and CO_2 gradient. Find **PMAX** (g $CO_2/m^2/hr$) if outside CO_2 concentration is 350 $\mu mol/mol$. The conversion factor at $0°C$ can be used.

14.   Compare the respiration function in this chapter with the function **RESP = Q10(TP)** shown in the section 10.2.

15.   Plot **LAI** as a function of **NL** by using eq. 10.23 with all values shown in the model in Fig. 10.8.

REFERENCES

ASHRAE, 1985: *ASHRAE Handbook* 1985 Fundamentals SI Edition.

Bautista, O. K., 1988: Use of plastics in Philippine agriculture. *Proc. of Int. Seminar on the Utilization of Plastics in Agriculture (ISUPA)*. Rural Development of Administration and Food & Fertilizer Technology Center.

Bennett, B. S., 1995: *Simulation Fundamentals*. Prentice Hall, 326pp.

Blom, Th. and Ingratta, F. J., 1985: The use of polyethylene film as greenhouse glazing in North America. *Acta Horticulturae*, **170**, 69-80.

Bualek, S., 1988: The Use of polymer materials in agriculture in Thailand. *ISUPA*.

Campbell, G. S., 1985: *Soil Physics with BASIC*. Transport models for soil-plant systems. Elsevier, 150pp.

Cekkkier, F. E., 1991: *Continuous System Modeling*. Springer-Verlag, 755pp.

De Wit, C. T. and van Keulen, H., 1972: *Simulation of Transport Processes in Soils*. PUDOC, Wageningen, 100pp.

De Wit, C. T. and Goudriaan, Y., 1974: *Simulaiton of Ecological Processes*. PUDOC, Wageningen, 175pp.

Fang, W., 1992. Theoretical investigation of solar radiation properties of multi-layer glazings – part I. Using ray tracing technique. *Journal of Agricultural Machinery, Taiwan*. 1(4) 31:41.

Fang, W., 1994. Limits and efficiencies of environmental control facilities in greenhouse of Taiwan (I). *Dept. of Agr. Mach. Eng., National Taiwan Univ., Research report*. NSC 83-0409-B-002-094. 173pp.

Forrester, J. W., 1971: *World Dynamics*. Wright-Allen Press, Cambridge Mass., 142pp.

Gates, D. M., 1962: *Energy Exchange in the Biolsphere*. Harper & Row Publ., 151pp.

Garnaud, J.-C., 1987: A survey of the development of plasticulture. Questions still to be answered. *Plasticulture*, 74, 5-14.

Grafiadellis, M., 1985: A study of greenhouse covering plastic sheets. *Acta Horticulturae*, **170**, 133-142.

Hanying Q. and Zhongdong, F., 1988: Agricultural films in China. *Plasticulture*, **77**(1), 25-28.

Hillel, D., 1980: *Application of Soil Physics*. Academic Press, 395pp.

Horie, T., 1987: The effect of climatic variations on rice yields in Hokkaido, in *The impact of climatic variations on agriculture*. Vol. 1, Assesments in cool temperature and cold regions, (ed. by Parry, M.L., Carter, T.R., and Konijn, N.T.), Reidel, Dordrecht, The Netherlands, 46-54.

IBM, 1975: *Continous System Modeling Program III* (CSMP III) Program Reference Manual Program Number 5734-XS9, SH19-7001-3, 206pp.

Ito, M., 1979: Studies on the gaseous environment in soil(4). The distribution of CO_2 concentraion in a soil and its diurnal change. *Environ. Control in Biol.*, **17**, 89-95.

Jansen, D. M., Dierkx, R. T., van Laar, H. H. and Alagos, M. J., 1988: *PCSMP on IBM PC-AT's or PC-XT's and compatibles*. Simulation Reports CABO-TT, No. 15, 64pp.

Japan Greenhouse Horticulture Association,1986: *Introductory Manual of Covering Materials in Protected Cultivation*. JGHA, 151pp.

Japan Greenhouse Horticulture Association (ed.), 2001: *Handbook of Protected Cultivation* (4th Edition). JGHA, 564pp.

Jensen, M. H., 1988: The achievements on the use of plastics in agriculture. *ISUPA*.

Jones, J. W., 1991: Grop growth, development, and production modeling. *Proc. of the 1991 Symposium on Automated Agriculture for the 21st Century*, ASAE, 447-457.

Jones, J. W., Dayan, E., Allen, L. H., van Keulen, H., and Challa, H., 1991: A dynamic tomato growth and yield model(TOMGRO). *Trans. ASAE*, 34, 663-672.

Jordan, K. A., 1992: Notes for ABE 415/515 Agri-biosystems Process Engineering. {Thermodynamics in Simulation of Biosystems}.

Jouët, J. - P., 2001: Plastics in the world. Plasticulture, **2**(120), 108-126.

Koorevaar, P., Menelik, G, and Dirksen C., 1983: *Elements of Soil Physics*. Developments in Soil Science 13, Elsevier, 230pp.

Kwon, Y. S., 1988: Improvement of vegetable cultivation by use of plastic films. *ISUPA*.

Martin, M. and Berdahl, P., 1984: Summary of results from the spectral and angular sky radiation measurement program. *Solar Energy*, **33**, 241-252.

Meadows, D. H., Meadows, D. L., Randers, J., and Behrens III, W. W., 1972: *The Limits to Growth*. Universe Books for Potomac Assoc., 205pp.

Monteith, J. L. and Unsworth, M. H., 1990: *Principles of Enviromental Physics* (2nd ed.) Edward Arnold. 291pp.

Nishi, S., 1986: *Protected Horticulutre in Japan*. Japan FAO Association 136pp.

Park, Y. D., 1988: The use of plastic films in agriculture. *ISUPA*.

Pegden, C. D.,1986: *Introduction to SIMAN* with Version 3.0 Enhancements. Systems Modeling Corporation.

Sellers, W. D., 1965: *Physical Climatology*. The Univ. of Chicago Press, 272pp.

Slatyer, R. O., 1967: *Plant Water Relationship*. Academic Press, 366pp.

Takahashi, K., 1975: *Environmental Control Techniques in Horticultural Engineering*. (Ed. by T. Takakura, et al.), Soft Science Co., 434pp.

Takakura, T. and Jordan, K. A., 1970: Comparison of simulation techniques in the prediction of plant leaf temperature and net photosynthetic rate in the greenhouse. *Proc. 1970 SCSC*, **2**, 779-785.

Takakura, T., Jordan, K. A. and Boyd, L. L., 1971: Dynamic simulation of plant growth and environment in the greenhouse. *Trans. ASAE*, **15**, 964-971.

Takakura, T., Goudriaan, J. and Louwerse, W., 1975: A behaviour model to simulate stomatal resistance. *Agric. Met.*, **15**, 393-404.

Takakura, T., Ohara, G. and Nakamura, Y., 1978: Direct digital control of plant growth. III. Analysis of the growth and development of tomato plants. *Acta Horticulturae*, **87**, 257-264.

Takakura, T.,1987: The use of agricultural plastics in Japan. *Proc. of the 20th Agricultural Plastics Congress*. Oregon, U.S.A., 253-257.

Takakura, T., 1988: Protected cultivation in Japan. *Acta Hort.*, 230, 29-37.

Takakura, T., 1989: Technical models of greenhouse environment. *Acta Hort.*, 248,

49-54.

Takakura, T., 1991: Environmental control systems for greenhouses. *Proc. of the 1991 Symposium on Automated Agriculture for the 21st Century*, ASAE, 437-446.

Takakura, T., Manning, T. O., Giacomelli, G. A. and Roberts, W. J., 1994: Feedforward control for a floor heat greenhouse. *Trans. ASAE* , 37, 939-945.

Ting, K. C. and G. A. Giacomelli. 1987. Availability of solar photosynthetically active radiation. *Tran. ASAE*, 30(5):1453-1456.

The Society of Agricultural Meteorology in Japan (SAMJ), Section for Protected Cultivation. 1986: *Proc. of National Symposium on Covering Techniques*. 21pp.

The Society of Agricultural Meteorology in Japan (SAMJ), Section for Protected Cultivation, 1987. *Present Situation and Discussion on Floating Mulches*. No.1. 56pp.

Threlkeld, J. L., 1962: *Thermal Environmental Engineering*. Prentice-Hall, Inc., 514pp.

Tokyo Astronomical Observatory (TAO) (ed.), 1986: *Chronological Scientific Tables*. 1017pp, Maruzen Co., Ltd.

Tzekleev, G., Guzelev, L., Solakov, Y. and Stoilov, S., 1988: Plasticulture in the People Republic of Bulgaria. *Plasticulture*, **78**(2), 19-23.

Van Bavel, C. H. M., 1951: A soil aeration theory based on diffusion. I. *J. Soil Science.*,**72**,33-46.

Van Keulen, 1975: *Simulation of Water Use and Herbage Growth in Arid Rigions*. PUDOC, Wageningen, 176pp.

Van Wijk, W. R. (ed.), 1966: *Physics of Plant Environment*. North-Holland Publ. Co., 382pp.

Wells, O. S., 1988: Improving crop production with row covers in the United States. *ISUPA*.

Yabuki, K., 1966: The carbon dioxide environment for plant, part 1. *Turf Research Bull.*, K.G.U. Green Section Research Center, **10**, 1-17.

APPENDIX

USER INTERFACE OF CUC MODELS

A.1. SET THE PATH

The programs mentioned in this book, entitled 'Climate Under Cover (CUC in short) , are available from the internet. The URL of the CUC web-site is listed in the 'Preface of the 2nd edition' of this book. All the CUC related programs, downloaded from the web-site, should be placed in one directory. The path to the directory should be assigned to MATLAB. One example is shown in Fig. A.1, which shows the path browser of **MATLAB** with a newly-added-path to a directory named 'C:\matlabR12\work\cuc-book'.

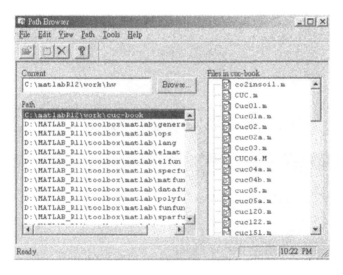

Figure A.1. Path browser of MATLAB v5.3 with the preset path to the directory containing CUC related files .

A.2. MENU AND SUBMENU OF USER INTERFACE

A '**cuc.m**' program was written to help the user in organizing all the models of the CUC book. By typing 'cuc' in the command window of **MATLAB** and followed by an 'Enter', a figure as shown in Fig. A.2 will pop up. This is the governing figure of all models. As shown in the second line from the top is the main menu containing options of 'Chap.3' to 'Chap.11' and 'Help'.

Fig. A.3 shows the submenu of 'Chap.3' option. In this submenu, there are 5 sub-options divided into 3 blocks. The first two will execute '**cuc01**' and '**cuc01a**' models. The third sub-option: 'close this' will close this main figure and the fourth sub-option 'close all' will close all the open figures. The last sub-option is to 'Quit Matlab'. Press Control key and Q key together is the quick exit to 'Quit Matlab'. This sub-option will close all figures and exit from **MATLAB**.

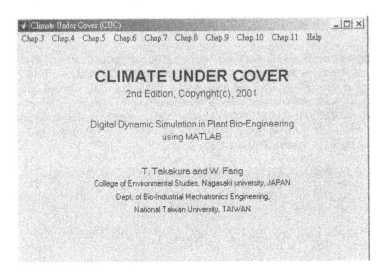

Figure A.2. Snapshot of the first figure of 'cuc.m' program.

Figure A.3. Submenu under Chap.3 option.

Fig. A.4 shows the submenu of 'Chap.4' and 'Chap.5' options.   Fig. A.5 shows the submenu of 'Chap.6' and 'Chap.7' options.   Fig. A.6 shows the submenu of 'Chap.8' and 'Chap.9' options.   Fig. A.7 shows the submenu of 'Chap.10' and Fig. A.8 shows the submenu of 'Chap.11' and 'Help' options.

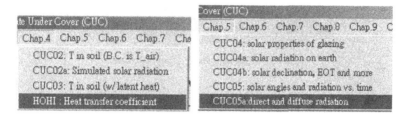

Figure A.4. Submenus under Chap.4 and Chap.5 options.

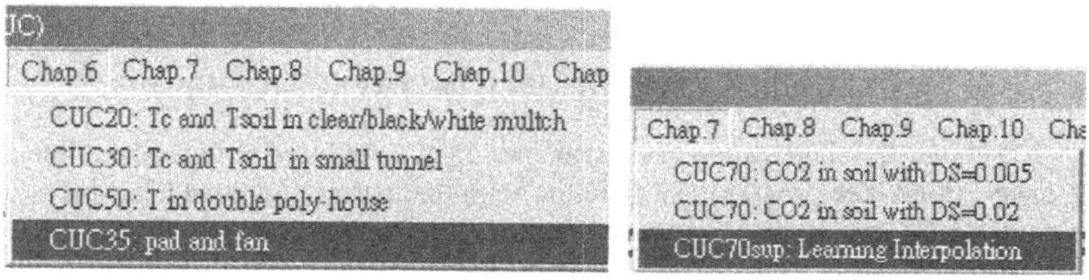

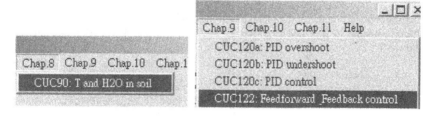

Figure A.5. Submenus under Chap.6 and Chap.7 options.

Figure A.6. Submenus under Chap.8 and Chap.9 options.

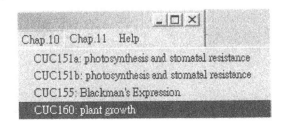

Figure A.7. Submenu under Chap.10 option.

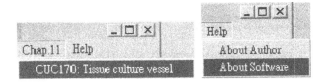

Figure A.8. Submenus under Chap.11 and Help options.

A.3. SOURCE CODE OF USER INTERFACE (**CUC**)

Fig. A.9 shows the source code of '**cuc.m**' program. As listed in the second line of the source code, **cuc** program is a function with one parameter named 'action'. If user enter only 'cuc' in the command window, the number of argument (**nargin**) is 0. In this case, the program will execute the 'init' portion within '**switch...case...end**'. There are 6 more cases besides 'init'. In this program, the powerful '**uimenu**' and '**uicontrol**' commands were used to create user interfaces as shown in Fig.A.2 to Fig.A.8.

A.3.1. User interface menu (**uimenu**)

Uimenu creates a menu on the menu bar at the top of the current figure window, and returns a handle to it. The first 3 lines after '%--[option1]--' in Fig. A.9 are as follows:

```
f1 = uimenu('Label', 'Chap.3');

uimenu(f1,'Label', 'CUC01: T in soil (B.C. is T_floor)', ...
                   'Callback','cuc01');
```

The left hand side of the first line is f1, which is the name of the 'handle'. The label of this handle is 'Chap.3' . The second line execute **uimenu** with f1 handle, which means it is the sub-option of 'Chap.3' menu and the label is 'CUC01: T in soil (B.C. is T_floor)'. When this sub-option is selected, the program will execute the value of 'Callback' property. In this case, the 'cuc01.m' program will be executed. The third **uimenu** of the same section is as follows:

```
uimenu(f1,'Label','Close this','Callback','cuc(''diabox'')', ...
                'Separator','on');
```

The value of 'Callback' property is 'cuc("diabox"). This means when 'Close this' sub-option is selected, the program will execute **cuc** program again with 'diabox' as the input parameter. This will execute the portion under

```
case 'diabox'.
```

A.3.2. question dialog box (*questdlg*)

In the 'diabox' section, the question dialog box is used. The scripts such as ButtonName = **questdlg** (question) creates a modal dialog box that automatically wraps the string question to fit an appropriately sized window. The name of the button that is pressed is returned in ButtonName. The default set of buttons names for **questdlg** are 'Yes','No' and 'Cancel'. The user can decide what to do next using '**switch..case...end**' function afterward.

A.3.3. User interface controller (*uicontrol*)

The user interface controller (**uicontrol**) creates a user interface control in the current figure window and returns a handle to it. Execute **get**(handle) to see a list of **uicontrol** object properties and their current values. In this program, the '**style**' property with the value 'text' and the value 'pushbutton' were used. Other properties and values please see online help of **MATLAB**.

```
% User interface for CUC models                              CUC.m
function cuc(action)
if nargin==0,  action='init'; end;
clc
switch (action)
case 'init'
h1=findobj('tag','CUC_MAIN');  close(h1);
figure('tag','CUC_MAIN','Resize','on','MenuBar','none',...
   'Name','Climate Under Cover (CUC)','NumberTitle','off',...
   'Position',[200,200,520,320],'color',[0.8 0.8 0.8]);
%--[option1]-------------------------------------------------
f1 = uimenu('Label','Chap.3');
uimenu(f1,'Label','CUC01: T in soil (B.C. is T_floor)', ...
   'Callback','cuc01');
uimenu(f1,'Label','CUC01a: T in soil (B.C. is T_floor)', ...
   'Callback','cuc01a');
uimenu(f1,'Label','Close this','Callback','cuc(''diabox'')', ...
    'Separator','on');
uimenu(f1,'Label','Close all','Callback','cuc(''CloseAll'')');
uimenu(f1,'Label','Quit Matlab','Callback','exit', ...
    'Separator','on','Accelerator','Q');
%--[option2]-------------------------------------------------
f2 = uimenu('Label','Chap.4');
uimenu(f2,'Label','CUC02: T in soil (B.C. is T_air)','Callback','cuc02');
uimenu(f2,'Label','CUC02a: Simulated solar radiation', ...
```

```
      'Callback','cuc02a');
uimenu(f2,'Label','CUC03: T in soil (w/ latent heat)', ...
      'Callback','cuc03');
uimenu(f2,'Label','HOHI : Heat transfer coefficient', ...
      'Callback','cuc(''runhohi'')');
%--[option3]-----------------------------------------------------------
f3 = uimenu('Label','Chap.5');
uimenu(f3,'Label','CUC04: solar properties of glazing', ...
      'Callback','cuc04');
uimenu(f3,'Label','CUC04a: solar radiation on earth', ...
      'Callback','cuc04a');
uimenu(f3,'Label','CUC04b: solar declination, EOT and more', ...
      'Callback','cuc04b');
uimenu(f3,'Label','CUC05: solar angles and radiation vs. time ', ...
      'Callback','cuc05');
uimenu(f3,'Label','CUC05a:direct and diffuse radiation ', ...
      'Callback','cuc05a');
%--[option4]-----------------------------------------------------------
f4 = uimenu('Label','Chap.6');
uimenu(f4,'Label','CUC20: Tc and Tsoil in clear/black/white multch', ...
      'Callback','test0(''cuc20'')');
uimenu(f4,'Label','CUC30: Tc and Tsoil  in small tunnel', ...
      'Callback','cuc30');
uimenu(f4,'Label','CUC50: T in double poly-house','Callback','cuc50');
uimenu(f4,'Label','CUC35: pad and fan','Callback','cuc35');
%--[option5]-----------------------------------------------------------
f5 = uimenu('Label','Chap.7');
uimenu(f5,'Label','CUC70: CO2 in soil with DS=0.005', ...
      'Callback','cuc70(1)');
uimenu(f5,'Label','CUC70: CO2 in soil with DS=0.02', ...
      'Callback','cuc70(2)');
uimenu(f5,'Label','CUC70sup: Learning Interpolation', ...
      'Callback','cuc70sup');
%--[option6]-----------------------------------------------------------
f6 = uimenu('Label','Chap.8');
uimenu(f6,'Label','CUC90: T and H2O in soil','Callback','cuc90');
%--[option7]-----------------------------------------------------------
f7 = uimenu('Label','Chap.9');
uimenu(f7,'Label','CUC120a: PID overshoot','Callback','cuc120(2)');
uimenu(f7,'Label','CUC120b: PID undershoot','Callback','cuc120(3)');
uimenu(f7,'Label','CUC120c: PID control','Callback','cuc120(1)');
uimenu(f7,'Label','CUC122: Feedforward & Feedback control', ...
      'Callback','cuc122');
%--[option8]-----------------------------------------------------------
f8 = uimenu('Label','Chap.10');
uimenu(f8,'Label','CUC151a: photosynthesis and stomatal resistance', ...
      'Callback','cuc151(1)');
uimenu(f8,'Label','CUC151b: photosynthesis and stomatal resistance', ...
      'Callback','cuc151(2)');
uimenu(f8,'Label','CUC155: Blackman''s Expression','Callback','cuc155');
uimenu(f8,'Label','CUC160: plant growth','Callback','cuc160');
%--[option9]-----------------------------------------------------------
f9 = uimenu('Label','Chap.11');
uimenu(f9,'Label','CUC170: Tissue culture vessel','Callback','cuc170');
%--[option10]----------------------------------------------------------
f10 = uimenu('Label','Help');
uimenu(f10,'Label','About Author','Callback','cuc(''author'')');
uimenu(f10,'Label','About Software','Callback','cuc(''version'')');
```

```
%--------------------------------------------------------------------------
maintxt1a='CLIMATE UNDER COVER';
maintxt1b='2nd Edition, Copyright(c), 2001';
maintxt2a='Digital Dynamic Simulation in Plant Bio-Engineering';
maintxt2b='using MATLAB';
maintxt3a='T. Takakura and W. Fang';
maintxt3b='College of Environmental Studies, Nagasaki university, JAPAN';
maintxt3c='Dept. of Bio-Industrial Mechatronics Engineering,';
maintxt3d='National Taiwan University, TAIWAN';
t1a = uicontrol('Units','normalized','Position',[.1, .8, .8, .1],...
    'string',maintxt1a,'style','text','fontname','Courier New', ...
    'fontSize',18,'FontWeight','Bold','backgroundcolor',[0.8 0.8 0.8]);
t1b = uicontrol('Units','normalized','Position',[.1, .75, .8, .05],...
    'string',maintxt1b,'style','text','fontname','Courier New',...
    'fontSize',10,'backgroundcolor',[0.8 0.8 0.8]);
t2a = uicontrol('Units','normalized','Position',[.1, .6, .8, .06],...
    'string',maintxt2a,'style','text','fontSize',12,...
    'backgroundcolor',[0.8 0.8 0.8]);
t2b = uicontrol('Units','normalized','Position',[.1, .54, .8, .06],....
    'string',maintxt2b,'style','text','fontSize',12,...
    'backgroundcolor',[0.8 0.8 0.8]);
t3a = uicontrol('Units','normalized','Position',[.1, .38, .8, .06],...
    'string',maintxt3a,'style','text','fontSize',12,...
    'backgroundcolor',[0.8 0.8 0.8]);
t3b = uicontrol('Units','normalized','Position',[.1, .32, .8, .06],...
    'string',maintxt3b,'style','text','fontSize',8,...
    'backgroundcolor',[0.8 0.8 0.8]);
t3c = uicontrol('Units','normalized','Position',[.1, .26, .8, .06],...
    'string',maintxt3c,'style','text','fontSize',8,...
    'backgroundcolor',[0.8 0.8 0.8]);
t3d = uicontrol('Units','normalized','Position',[.1, .2, .8, .06],...
    'string',maintxt3d,'style','text','fontSize',8,...
    'backgroundcolor',[0.8 0.8 0.8]);
%--------------------------------------------------------------------------
case 'cuc20'
   cuc20(1);    cuc20(2);    cuc20(3);
%--------------------------------------------------------------------------
case 'diabox'
   selection = questdlg('Close Current Figure?', 'What''s next?',...
                'Yes','No','Cancel');
   switch selection,
    case 'Yes',
        delete(gcf);
    case {'No','Cancel'}
        return
    end
%--------------------------------------------------------------------------
case 'runhohi'
   figure;    hohi; % run hohi.m
%--------------------------------------------------------------------------
case 'CloseAll'
   close all;
%--------------------------------------------------------------------------
case 'version'
    figure('tag','INPUT','Resize','off','MenuBar','none','Name', ...
        'Version 2.0','NumberTitle','off','Position',[140,200,200,60]);
    t1 = uicontrol('Units','normalized','Position',[.1, .1, .8, .8],...
        'string','Updated: 2002/3/12.','style','text');
```

```
   act1 = uicontrol('Units','normalized','Position',[.4,.1, .2, .3],...
   'string','O.K.', 'style','pushbutton','callback','close');
%---------------------------------------------------------------------
case 'author'
   figure('tag','INPUT','Resize','off','MenuBar','none',  ...
       'Name','Program written by ',  ...
       'NumberTitle','off','Position',[140,200,250,100]);
   txt1='Professor Wei Fang, Ph.D., ';
   txt2='Dept. of Bio-Industrial Mechatronics Engineering, ';
   txt3='National Taiwan University, Email: weifang@ccms.ntu.edu.tw';
   showtxt=[txt1 txt2 txt3];
   t1 = uicontrol('Units','normalized','Position',[.1, .1, .8,.8],...
       'string',showtxt,'style','text');
   act1 = uicontrol('Units','normalized','Position',[.4,.1,.2,.2],...
   'string','O.K.','style','pushbutton',...
   'callback','close');
end
```

*Figure A.9. Scripts of user interface program (**cuc.m**).*

INDEX